Cybersecurity Blue Team Stra

Uncover the secrets of blue teams to combat cyber threats in your organization

Kunal Sehgal

Nikolaos Thymianis

BIRMINGHAM—MUMBAI

Cybersecurity Blue Team Strategies

Group Product Manager: Mohd Riyan Khan
Publishing Product Manager: Mohd Riyan Khan
Content Development Editor: Nihar Kapadia
Technical Editor: Arjun Varma
Copy Editor: Safis Editing
Project Coordinator: Ashwin Dinesh Kharwa
Proofreader: Safis Editing
Indexer: Sejal Dsilva
Production Designer: Shyam Sundar Korumilli
Senior Marketing Coordinator: Marylou De Mello

First published: February 2023

Production reference: 2270123

Published by Packt Publishing Ltd.
Livery Place
35 Livery Street
Birmingham
B3 2PB, UK.

ISBN 978-1-80107-247-2

www.packtpub.com

For my grandfather, who showed me what it means to be truly devoted to one's work.

– Kunal Sehgal

To my father, Stefanos Thymianis, and to my mother, Vasiliki, for all their sacrifices and for exemplifying the power of determination. To my brother, Mario, for always being there to make jokes and offer exemplary guidance when needed.

– Nikolaos Thymianis

Contributors

About the authors

Kunal Sehgal has been a cyber-evangelist for over 15 years and is an untiring advocate of Cyber Threat Intelligence sharing. He encourages cyber-defenders to work together by maintaining a strong level of camaraderie across public and private sector organizations. He has worked on setting up two Information Sharing & Analysis Centers to combat cybercrime, and regularly shares credible intelligence with law enforcement agencies around the world.

Kunal has also worked for various organizations, in leadership roles, to drive security improvement initiatives and to build cybersecurity services, especially within the APAC region. He specializes in helping businesses improve their security posture and resilience while leveraging the power of the cloud.

Kunal resides in Singapore, and invests his non-working hours in researching, blogging, and presenting at cyber-events across Asia. He has 17 certifications/degrees in various IT- and information security-related topics.

This book would not have been possible without the support of my wife and the numerous times I had to cancel plans with her just to work on this. This book is dedicated to my kids, who drive me to be a better person, and my brother and mother, who have always been my pillars of support.

Nikolaos (Nick) Thymianis studied cultural informatics at the University of the Aegean in Greece, during which he received a scholarship to go to the UK and continue his education to gain an MSc in information security, at the University of Brighton.

Nick's work experience led him to associate with people in the healthcare industry, while doing cybersecurity assurance and maturity assessments for organizations in the NHS, helping to set the standards and guidelines for hospitals in the UK. Nikolaos was the CISO of caresocius from 2018 until 2022. Nick is now active in big pharma, working in risk management/exception management. He always encourages everyone he meets to be security aware, because information security is a problem everyone faces.

He is an advisor at the University of Piraeus and has also become a recognized cybersecurity speaker.

I want to thank the people who have been close to me and supported me, especially my family and friends who, even when I thought of giving up, always raised me up.

About the reviewers

Stefano Barber is a graduate of RheinMain University with a Bachelor of Science in business informatics.

His education and project management experience spans various industries, including the automotive industry, technology, and healthcare. His areas of expertise include data protection consultancy, AI research, and the implementation of enterprise-wide systems.

As an entrepreneur and advocate for digital solutions, Stefano is the cofounder and head of operations at caresocius. The Swiss healthcare company was established in 2018 and under his leadership has grown into a secure online platform for the referral of patients to international clinics and medical service providers.

Avinash Sinha is a Cyber Security Expert with 12+ years of experience & has worked with many Global Business Unicorns and MNC's such as IBM, GE HealthCare, Schneider Electric, Airtel, Target, Aujas Networks, UIDAI Aadhaar, securing them from Next Generation threats.

He has Managed & Lead teams for successful execution of OT/IT/Cloud Security projects entailing Security Operations, Enterprise Security Architecture, Penetration Testing, HIPAA, GDPR, Threat Intelligence, IR, Red Teaming, Cloud Security assessment for O365, Azure, Citrix & AWS.

He has Degree in Computer Technology with majors in Artificial Intelligence and PG in International Business from Symbiosis. He holds GICSP GIAC cert ,GCFA, SEC 541-Cloud Attacks and Monitoring from SANS.

I'd like to thank my father Rajendra Sinha for constantly inspiring me and my Mother for her Love and blessings. Also thanks to my lovely Wife Tanu for taking care of me and keeping me healthy while my passion for Cyber security extends beyond unusual times.

Dedicated to my son, Kiyansh Sinha, you are always in our prayers. Thanks for bringing joy and happiness into our lives. I hope your next life is full of happiness and bliss.

Table of Contents

Preface   xiii

Part 1: Establishing the Blue

1

Establishing a Defense Program   3

How do organizations benefit from
implementing the blue teaming
approach?   4
Risk assessment   4
Monitoring and surveillance   4
Security controls   4
Reporting and recommendation to management   5

A blue team's composition   5
Analysts   5
Incident responder   6
Threat hunter   6
Security consultant   7
Security administrator   7
Identity and Access Management (IAM)
administrator   8
Compliance analyst   8

Red team   8

Purple team   9
Cyber threat intelligence   10
Skills required to be in a blue team   10
Eager to learn and detail-oriented   11
In-depth knowledge of networks and systems   11
Outside-the-box and innovative thinking   11
Ability to cross conventional barriers to
perform tasks   11
Academics, qualifications, and certifications   12

Talent development and retention   12
Cyber labs   12
Capture-the-Flag and hackathons   12
Research and development projects   13
Community outreach   13
Mentoring   13
Continuous unhindered learning   13

Summary   14

2

Managing a Defense Security Team   15

**Why must organizations consider
metricizing cybersecurity?**   15
Blue team KRIs   16
How does a blue team initiate designing
KRIs for their team?   17
Selecting essential cybersecurity metrics   19

**Why and how organizations
can automate this process**   22
What pitfalls to avoid when automating
the workflows of the blue team   22
Automating how KRIs are collected
and presented   23

Summary   24

3

Risk Assessment   25

Following the NIST methodology   25
NIST risk assessment methodology   27

Asset inventory   28
Risk management methods   31
Threat identification   31

Risk calculation   33

Risk management responsibilities   36
Summary   37
References   37

4

Blue Team Operations   41

Understanding defense strategy   41
Blue team operations – infrastructure   43
Blue team operations – applications   44
Blue team operations – systems   46
Blue team operations – endpoints   47

Blue team operations – cloud   48
Defense planning against insiders   51
**Responsibilities in blue team
operations**   53
Summary   54

5

Threats   55

What are cyber threats?   55
The Cyber Kill Chain   57

Phase 1 – reconnaissance   57
Phase 2 – weaponization   60
Phase 3 – delivery   62
Phase 4 – exploitation   62
Phase 5 – installation   63
Phase 6 – command and control   63
Phase 7 – actions on objective   64

Internal attacks   70
Different types of cyber threat actors   71
Impacts of cybercrime   72
An approach to security that is proactive
rather than reactive   73
Summary   73

6

Governance, Compliance, Regulations, and Best Practices   75

Definition of stakeholders
and their needs   75
Building risk indicators   77

Compliance needs and the
identification of compliance
requirements   79
Assurance of compliance and the
right level of governance   83
Summary   85

Part 2: Controlling the Fray

What are security controls?   87
Preventive controls   88
Detective controls   88
Deterrent controls   89

Compensating controls   89
Corrective controls   89
Defense-in-depth   90

7

Preventive Controls   93

What are preventive controls?   93
Benefits   93

Types of preventive controls   94
Administrative   94
Physical   95
Technical/logical   95

Layers of preventive controls   96
Policy control   96

Perimeter/physical controls   98
Network controls   99
Data security controls   100
Application security controls   101
Endpoint security controls   102
User security   103

Summary   104

8

Detective Controls   105

What are detective controls?   105   Source code scanning   112
Types of detective   Compliance scanning or
controls   106   hardening scans   112
SOC   107   Tools for detective controls   113
How does a SOC work?   107   Threat Intelligence Platform (TIP)   113
What are the benefits of a SOC?   108   Security Orchestration, Automation, and
Vulnerability testing   109   Response (SOAR) tools   114
                                           Security Information and Event
Penetration testing   110   Management (SIEM) tools   114
Red teams   111   Digital Forensics (DF) tools   115
Bug bounty   111   Summary   116

9

Cyber Threat Intelligence   117

What is CTI?   117   4 – Analysis   123
The quality of CTI   118   5 – Dissemination   123
                                           6 – Feedback   124
Types of threat intelligence   119   Threat hunting   124
Strategic threat intelligence   119   The importance of threat hunting 124
Tactical threat intelligence   120   Using CTI effectively   126
Operational threat intelligence   121   The MITRE ATT&CK framework   127
Threat intelligence implementation 122   The MITRE ATT&CK Matrix   127
1 – Developing a plan   122   How to implement the ATT&CK framework   128
2 – Collection   122   Summary   129
3 – Processing   123

10

Incident Response and Recovery   131

Incident response planning   131   Incident response playbooks   136
Testing incident response plans   133   Ransomware attacks Playbook   136

Data loss/theft attacks playbook   141

Phishing attacks playbook   145

Disaster recovery planning   149

Cyber insurance   153

Summary   155

11

Prioritizing and Implementing a Blue Team Strategy   157

Emerging detection and prevention technologies and techniques   158

Adversary emulation   158

VCISO services   159

Context-aware security   159

Defensive AI   160

Extended Detection and Response (XDR)   160

Manufacturer Usage Description (MUD)   160

Zero Trust   161

Pitfalls to avoid while setting up a blue team   162

Getting started on your blue team journey   164

Summary   164

Part 3: Ask the Experts

12

Expert Insights   169

Anthony Desvernois   169

William B. Nelson   170

Career   170

Non-profit and volunteer work   171

Laurent Gerardin   172

Peter Sheppard, BSc (Hons), MBCS, CITP, CISA   173

Pieter Danhieux, CEO and Co-Founder, Secure Code Warrior   174

Index   177

Other Books You May Enjoy   188

Preface

Setting up a cyber blue team may seem like a daunting task. There are a lot of considerations to keep in mind, and the list of products and services on the market is endless. It is easy to get lost in a sea of jargon and lose sight of what is absolutely needed for an organization. This book is meant for professionals looking to get started on building such a capability. The primary intention of this book is to guide you along each step of the journey, explaining what any organization should consider, and how to design a comprehensive defense capability.

Disclaimer

The opinions expressed in this publication are those of the authors alone. They do not purport to reflect the opinions or views of our employers, if any, or of any third parties. Any designations used, any views expressed, the presentation of materials, and the use of references in this publication do not constitute and do not imply the expression of any opinion whatsoever by, or any involvement of, any employer legal entity that either of the writers may have. Our employers, if any, make no representation and assume no liability whatsoever for the accuracy of the information contained in this publication and for the views expressed by the undersigned authors of the present publication.

Nikolaos Thymianis   Kunal Sehgal

Nikolaos Thymianis   *Kunal Sehgal*

Who this book is for

This book is meant for anyone looking to embark on the journey of setting up a cyber defense team (AKA a blue team) for their organization. It is business agnostic, and hence professionals from all fields will find it equally useful. The primary goal of the book is to explain all the aspects of setting up such a capability and to ensure there is comprehensive coverage and no blind spots. This involves understanding not just the organization's needs and risk appetite, but also looking into the applicable laws and regulations, before designing the relevant controls. This will ensure the organization gets the most value from its investments, and that the designed defense capabilities are fit for purpose.

The book is designed to keep senior executives in mind. Hence, **Chief Information Security Officers (CISO)**, **Chief Information Officers (CIOs)**, board members, and other C-level executives will benefit from the strategies and concepts introduced in this book. However, even junior professionals in the information security domain will find value in collecting their thoughts to design a plan that could be presented to the senior executives at their respective organizations.

What this book covers

Chapter 1, Establishing a Defense Program, provides a general description of what a blue team is and what its role is in the business. Moreover, it contains a historic review of how blue teams came to be. It also discusses the difference between red teams and blue teams.

Chapter 2, Managing a Defense Security Team, explains the role this team should play in an organization, and also what processes to build up and what responsibilities to give to such a team. Moreover, the chapter discusses how this team would work with the other departments in an organization.

Chapter 3, Risk Assessment, explores risk assessments, how a blue team should go about conducting one, and how to calculate risk for their organization.

Chapter 4, Blue Team Operations, explores the blue team operations that should be considered by an organization when they consider setting up cyber defence capabilities, including what key focus areas to look into and how to avoid any blind spots.

Chapter 5, Threats, explores how a blue team should go about identifying the major threats to their organization, that is, how to classify, assess, and prioritize risks.

Chapter 6, Governance, Compliance, Regulations, and Best Practices, explains what governance is, how to do it correctly, and how to provide visibility to all the stakeholders in the organization. You will also learn why it is important to be aware of any external requirements, to ensure they are based at the right level, and lastly, what to expect from major regulations (such as GDPR).

Chapter 7, Preventive Controls, covers the various controls that a defense team should consider. The chapter is structured as per the NIST framework, which will be touched upon briefly. The intention is to help you understand the full spectrum of controls to consider.

Chapter 8, Detective Controls, goes through why detective controls are needed and how to augment preventive controls. Moreover, the chapter reviews how such controls work in a typical organization, and what processes are needed in tandem with the technology to ensure an adequate level of security.

Chapter 9, Cyber Threat Intelligence, delves into threat intelligence, its foundation, and how it is an important tool in the arsenal of a blue team. Secondly, the chapter explains how a blue team can keep itself updated on the latest threats and methods.

Chapter 10, Incident Response and Recovery, explains how to make incident response plans, how to test those plans, and what to do about cyber-insurance. The chapter also covers the NIST: Respond, Recover methodology and explains it thoroughly with examples from incident response teams.

Chapter 11, Prioritizing and Implementing a Blue Team Strategy, summarizes everything we have learned in this book, and how to prioritize various steps to suit your organization. This chapter also refers to emerging technologies and methodologies that are becoming commonplace in the information security industry.

Chapter 12, *Expert Insights*, introduces industry experts who will share their views on the book. They will share from their own experience how they went about establishing their own blue-team processes and what tools or frameworks helped them along the way.

To get the most out of this book

Some basic knowledge of information security is recommended, but not mandatory.

Download the color images

We also provide a PDF file that has color images of the screenshots and diagrams used in this book. You can download it here: `https://packt.link/Vke8t`.

Conventions used

There are a number of text conventions used throughout this book.

> **Tips or important notes**
> Appear like this.

Get in touch

Feedback from our readers is always welcome.

General feedback: If you have questions about any aspect of this book, email us at `customercare@packtpub.com` and mention the book title in the subject of your message.

Errata: Although we have taken every care to ensure the accuracy of our content, mistakes do happen. If you have found a mistake in this book, we would be grateful if you would report this to us. Please visit `www.packtpub.com/support/errata` and fill in the form.

Piracy: If you come across any illegal copies of our works in any form on the internet, we would be grateful if you would provide us with the location address or website name. Please contact us at `copyright@packt.com` with a link to the material.

If you are interested in becoming an author: If there is a topic that you have expertise in and you are interested in either writing or contributing to a book, please visit `authors.packtpub.com`.

Share your thoughts

Once you've read *Cybersecurity Blue Team Strategies*, we'd love to hear your thoughts! Scan the following QR code to go straight to the Amazon review page for this book and share your feedback.

https://packt.link/r/1-801-07247-7

Your review is important to us and the tech community and will help us make sure we're delivering excellent quality content.

Download a free PDF copy of this book

Thanks for purchasing this book!

Do you like to read on the go but are unable to carry your print books everywhere? Is your e-book purchase not compatible with the device of your choice?

Don't worry! With every Packt book, you now get a DRM-free PDF version of that book at no cost.

Read anywhere, any place, on any device. Search, copy, and paste code from your favorite technical books directly into your application.

The perks don't stop there! You can get exclusive access to discounts, newsletters, and great free content in your inbox daily.

Follow these simple steps to get the benefits:

1. Scan the QR code or visit the link:

https://packt.link/free-ebook/9781801072472

2. Submit your proof of purchase.
3. That's it! We'll send your free PDF and other benefits to your email directly.

Part 1: Establishing the Blue

In this part, you will learn what a blue team is and how they work. You will learn about the history of how the blue team came to be and what this means for general business. We will also mention different fields where defense plays an important role, such as cloud security, network security, and so on. Moreover you will also learn how to recruit and build a blue team.

This part of the book comprises the following chapters:

- *Chapter 1, Establishing a Defense Program*
- *Chapter 2, Managing a Defense Security Team*
- *Chapter 3, Risk Assessment*
- *Chapter 4, Blue Team Operations*
- *Chapter 5, Threats*
- *Chapter 6, Governance, Compliance, Regulations, and Best Practices*

Establishing a Defense Program

As cyberattacks ramp up across all countries and industries, it is an absolute necessity for every organization to have a defense capability. However, the journey of setting up such expertise and attaining the right level of maturity requires the right combination of technology, processes, and people. This roadmap may appear daunting and overwhelming to many who are just getting started. This book aims to help guide and aid organizations and professionals on that journey. It aims to ensure all aspects of a blue team defense program are understood and that there are no blind spots.

Cybersecurity professionals who are grouped under the banner of *blue team* identify various security holes, also known as vulnerabilities, in the organization's infrastructure and applications. These efforts help in patching and implementing various security procedures and controls. Cyber professionals working as blue teamers usually have a knack for creatively thinking and rapidly responding to various kinds of security events and incidents. They are in charge of protecting business entities against cyber risks and threats.

In this chapter, we will cover the following topics:

- How do organizations benefit from implementing the blue teaming approach?
- A blue team's composition
- Red team
- Purple team
- Cyber threat intelligence
- Skills required to be in a blue team
- Talent development and retention

How do organizations benefit from implementing the blue teaming approach?

Before we start, it is important to understand the benefits an organization can expect to achieve from setting up a blue team. This chapter will focus on what an organization can expect to gain from setting up a blue team, and how to take step-by-step action to set one up for success.

Risk assessment

First, businesses are recommended to assess the risks and threats that affect their organizational assets located across the globe. Blue teamers perform a risk assessment to learn how and what is to be defended from cyberattacks. They typically recommend implementing stringent security controls and establishing standard procedures to improve the security posture of the organization. Often, they design the structure of the **End User Security Awareness** training as well. This helps organizations identify their critical assets and the threat profile for each asset and the organization as a whole.

Monitoring and surveillance

Monitoring and surveillance are the core tasks of blue teamers; they perform them diligently for their respective businesses. Organizations receive recommendations for procuring, deploying, and launching various security monitoring tools from blue teamers. These tools allow organizations to log information about the various kinds of access privileges that the users and employees have on the network infrastructure. All the user activities are recorded, and suspicious activities trigger alerts as per the rules configured in the various security tools. Daily checks such as auditing DNS and firewall configuration, performing daily compliance checks across the dashboards of different tools deployed, and others are some of the **Key Responsibility Areas (KRAs)** of blue teamers. They also perform various kinds of internal and/or external vulnerability assessments on the network. Once in a while, blue teamers help prioritize and provide guidance to patch the vulnerabilities discovered in the penetration test reports. Blue teamers are experts in scanning the business network for vulnerabilities as well as analyzing the captured network packets for suspicious ingress and/or egress traffic.

Security controls

Blue teamers are also tasked with establishing various kinds of technical security controls over critical assets. Hence, they have to identify and classify the most critical network components in the organization. Organizations can utilize a **Configuration Management Database (CMDB)** to document the change in any configuration they make to those assets. Also, CMDBs are used to centralize a record of all the network components in any network infrastructure. Assets that are likely to shut down the business altogether if they are hit by cyberattacks are categorized as *critical assets*. Most of these assets are hardened with additional security controls. Along with risk assessment, blue teamers perform impact assessment studies as well. This involves calculating the impact that various cyberattacks could have if they hit certain critical assets and if those assets go down for a specific time. This could seriously affect business operations on a large scale. Hence, the risks and threats that affect every asset that falls under the critical category are documented. Regular vulnerability assessment scans are scheduled for all the disclosed vulnerabilities that affect those assets – namely **Common Vulnerabilities and Exposures (CVEs)** and **Common Weakness Enumeration (CWE)**. Blue teamers are proficient at assessing risks and suggesting remediation steps for them as well. Most of the critical and high-level vulnerabilities are patched as soon as possible. A plan is put into action by the blue teamers so that they can implement the security controls that eventually aim to decrease the impact of those vulnerabilities for which patches haven't been released yet.

Reporting and recommendation to management

The executive team must decide whether the security controls that are in place are adequate. Blue teamers prepare a document of the known risks that the business is running. Blue teamers may also perform cost-benefit analysis for management to recommend only those security controls deemed crucial to be implemented on a bare-minimum basis.

As an example, blue teamers may discover that the company's network is vulnerable to **Distributed Denial-of-Service (DDoS)** attacks. DDoS attacks deny the network's availability to genuine users by flooding traffic requests to the company servers. Here, the unavailability of services might result in revenue losses for the business. The more time it takes the network team to block a certain subnet of IP addresses, the more losses the business encounters. These kinds of attacks severely cripple the organizational network. Here, the blue team not just analyzes and tries to help in blocking the C2 IP addresses of the attackers but also performs impact assessments. To prevent DDoS or any type of **Denial-of-Service (DoS)** attack, blue teamers recommend deploying perimeter security solutions. These software solutions drastically lower the likelihood of the organization being affected by DDoS attacks. They do not and cannot stop one from originating, but they can certainly stop one from affecting your business network. Security solutions such as perimeter firewalls, load balancers, and WAF help in detecting DoS attacks and preventing them from affecting your organizational network.

There are many other advantages of setting up a blue team; this section only provided an overview of what the typical advantages are. Next, we will focus on what skills and talent to hire in such a team.

A blue team's composition

A blue team comprises many individuals with diverse skill sets. The composition of a team differs per the needs of an organization. In this section, we'll look at a few typical roles that usually sit within this team.

Analysts

An entry-level cybersecurity role known as *SOC analyst* exists in the company's **Security Operations Center (SOC)**. A cybersecurity analyst is also known as a triaging analyst. The SOC analyst responds to specific severity incident alerts and investigates the evidence. This role is reactive. Organizations usually have **Level 1 (L1)**, **Level 2 (L2)**, and **Level 3 (L3)** roles in SOC. L1 is the most junior analyst role, whereas L3 is the senior-most analyst role in a SOC. In most cases, the rising numbered levels are utilized to denote increasing levels of responsibility and experience requirements.

SOC monitors IT network traffic for unusual or suspicious behavior. Certain suspicious activities might indicate the existence of malicious entities or malicious programs such as Trojans and ransomware in the network. Senior analysts examine the alerts generated by the **Security Incident and Event Management (SIEM)** solution (such as Splunk, IBM QRadar, Logrhythm, and others). Analysts work on triaging and identifying suspicious events and determine whether the alerts are false positives or true positives. In the case of true positive alerts, the predefined **Standard Operating Procedure (SOP)** according to the playbooks or runbooks is followed. The analysis and investigation that are performed by the junior analysts help establish a context for the security incidents that have occurred. They also determine the severity of a security issue and apply the appropriate risk rating to it. Security incidents with critical and high severity are immediately escalated to the **Incident Responder (IR)** in the SOC team.

Incident responder

An IR is also known as an incident response analyst. This position assesses if a reported alarm is an organizational attack or a persistent danger to a company's network. They ensure that it is contained as quickly as possible and that the organization can respond and recover from it as per the defined plans. IRs usually investigate the scope of a cyberattack.

Based on the extent of the cybersecurity problem, IRs devise a remediation strategy. This entails investigating the incident's characteristics. This includes the business assets targeted by malware as well as the types of harmful activities performed by the malware. Then, the IRs recommend the appropriate course of action. They implement remediation with the necessary teams, such as initiating IT tickets to re-image compromised systems. Often, IRs face the heat of pushing the essentiality of mandating end user security awareness training by the CISO. They also notify the chief executives of the scope of a data breach in a timely way.

Threat hunter

Often, this work role is known as *threat analyst* or *threat researcher*. The threat hunter's work is *proactive*. They regularly research threats and risks to keep themselves updated on the newest threats. They also study the evolution and anatomy of threats. Threat hunters often design coding rules that trigger alerts in the company's SIEM solution for specific cyber threats.

Threat hunters are proficient in configuring as well as monitoring multiple threat intelligence platforms (for example, IBM X-Force, Alienvalult OTX, VirusTotal, and others) to conduct proactive research into the threats' life cycle. They assess whether new and emerging threats provide the most danger to their company based on various parameters, such as the industries targeted, vulnerabilities exploited, and attack TTPs. Threat hunters often implement system configuration adjustments to respond to the cyber risks that have been discovered. Analyzing cyber threats and risks in real time becomes overwhelming when the threat intelligence that's received is more than what the human resources provided can process. Hence, threat hunters use automation in security technologies to detect behavior that is typical of certain threats automatically. They sensitize and strengthen the organizational network infrastructure to withstand potential cyberattacks.

Let's presume that a novel ransomware cyber threat has surfaced recently (such as **Lockbit 2.0** or **BlackMatter**). A threat hunter will investigate this danger and use automation to prevent it from infiltrating the company and identify it if it does intrude.

A candidate is required to be experienced in the SOC analyst and IR work roles as well as proficient in computer and systems networking and administration to get hired for a threat hunter role. Also, it is good to be familiar with the various sources of threat intelligence on the surface of the web, as well as the dark web. Having a deep understanding of the business sector-specific cyber threats often provides the candidate with a competitive edge in the threat intelligence and threat hunting job market. A good threat hunter or **Threat Intelligence Analyst** (**TIA**) is proficient in obtaining proactive and actionable **Threat Intelligence** (**TI**) via any number of sources from the surface of the web, as well as the dark web, including the various **Internet Relay Chat** (**IRC**) servers and forums. A good threat hunter must be able to choose the appropriate technical and non-technical methodologies, as well as have the know-how to use various TI frameworks at their disposal.

Security consultant

Security consultants are often hired on a contractual basis and perform tasks throughout the project's life cycle as and when required. They may also be hired from outside the organization to bring in a reliable source of knowledge or expertise in a specific tool or area of security. They are often regarded as experts in their domain of knowledge. Another term often used to designate security consultants is **Subject Matter Experts** (**SMEs**). Security strategy consultant and security operations consultant are a few examples of specialized roles.

Security administrator

A security administrator is not the same as a SOC analyst. However, often, it has been seen that organizations consider security administrators as **Level 4** (**L4**) SOC analysts, whose job is to download, install, configure, deploy, and launch various security tools in the SOC. They also take care of updating those tools when the vendor updates arrive. This job is similar to that of a systems administrator's, but it deals with all the security tools in SOC such as SIEM, SOAR, AV-NGAV, EDR-XDR, DLP, honeypots, cloud governance, WAF, firewall, load balancers, IAM and AD, brand abuse and defamation monitoring solutions, and more. The job also entails applying patches or fixes released by the respective tools' vendors and configuring security tools to ensure optimum performance. They often collaborate with threat hunters and IRs to create security scripts and programs that automate some of the redundant security tasks. However, they are not tasked with investigating security events and incidents flagged by the security tools.

Identity and Access Management (IAM) administrator

This role provides **Identity and Access Management (IAM)** support to several departments within a firm. Managing application/system authority and privileges, **Single Sign-On (SSO)**, application reporting, and working with developers to integrate identity and access management policies for new applications and software are some of the key responsibilities of an IAM admin. These professionals have niche expertise in the use of various IAM tools, as well as networking administration.

Compliance analyst

A compliance analyst is often tasked with the internal audits of a corporation or a business. They check and verify whether the business is following its security rules, privacy policies, national data privacy laws, or any other applicable laws/regulations. They have experience in all the aforementioned work roles since a compliance analyst is required to handle frequent discussions with all the other work roles as part of compliance checks. They derive regular reports of non-compliance found or detected in the network infrastructure and submit them to senior management. Additionally, they assist firms in preparing for external audits, which may be necessary, depending on the business sector (for example, healthcare, BFSI, energy and utilities, and others).

This section covered what organizations need to understand to compose a blue team. There will be more roles to consider, depending on the type or complexity of an organization. However, in this section, we covered some of the skills that are typical in any organization. Next, we will briefly touch upon the red team and the purple team. These two teams may not be part of a blue team, but it is important to understand what these teams do as well. Moreover, we will also understand the role of a cyber threat intelligence team. This skill set typically sits within the blue team, but it is also common to have this team segregated from the blue team.

Red team

The red team behaves like hackers who attempt to find and exploit any potential loopholes inside a business network. Red teamers are known to use a wide range of conventional as well as unconventional techniques to uncover flaws in technology, people, and processes. Hence, usually, such a skill set would exist outside the scope of that of a blue team. However, for the sake of understanding, let's briefly touch upon this role.

A red team's mission consists of searching for known vulnerabilities that have already been disclosed and have a **Common Vulnerabilities and Exposures (CVE)** ID. They perform penetration tests on the business network infrastructure to discover unknown security loopholes. These teams may also test the wireless and IoT networks, along with the endpoint devices, such as laptops, PCs, mobiles, tablets, and more. Hardware penetration testing is performed on IoT wearables and devices that utilize Bluetooth. The hackers in red teams may try to social engineer the employees of their organization. These kinds of hackers are often assigned aliases to operate on the company's premises. They are very crucial in detecting as well as suggesting the security controls required to patch the security breaches that occur through a lack of physical measures in place. Endpoints and mobile devices are also covered in their scope of penetration or intrusion tests.

The detailed responsibilities of a red team are beyond the scope of this chapter. However, it is important to note that typically, the red team and the blue team work in tandem. Some of the areas where they work together are as follows:

- Creating a network topology/hierarchy map of the business's network infrastructure so that they can analyze the number of hosts running, as well as their statuses

- Assessing the services running and the open ports on those systems

- Identifying the vendor, firmware, and OS details among other relevant equipment parameters

- Identifying and exploiting the CVEs in servers, hubs, firewalls, routers, L2/L3 switches, Wi-Fi access points, and other network equipment

- Hacking various kinds of physical security controls, such as glass doors, digital locks, CCTV networks, and sometimes the security personnel as well

In some organizations, it may also be wise to set up a bug bounty program. A bug bounty is either a sum of money or goodies paid or provided to ethical hackers. Hackers throughout the world are on the lookout for defects and, in some circumstances, make a living doing so. Many websites, organizations, and software companies provide bug bounty programs in which users can be recognized and compensated for reporting bugs, particularly those related to business logic vulnerabilities and network security exploits. Bug bounties are created by companies to reward independent bug bounty hunters who find security flaws and weaknesses in systems. Companies pay bounty hunters to find security flaws and report them ethically and responsibly before they can be exploited or monetized by cyber threat actors. Bounty programs are frequently used in conjunction with regular penetration testing to allow enterprises to assess the security of their apps throughout their development life cycle. Bug bounty schemes enable businesses to use the hacker community to continually enhance the security posture of their systems. Bounty schemes attract a diverse group of hackers with various skill sets and expertise, offering firms an advantage over vulnerability assessments, which rely on inexperienced security personnel. Hence, instead of one individual or one team working on attacking the defenses of an organization, the collective power of the crowd benefits the organization.

Purple team

The fundamental goal of the red and blue team exercises is to improve the organization's overall security posture. This is where the purple team notion comes into play. A purple team isn't always a standalone group, though it may be. A purple team's purpose is to bring together the red and blue teams while encouraging them to collaborate and exchange ideas to build a strong feedback loop. The purpose of a purple team is to develop blue team capabilities while maximizing the results of red team engagements. A company functions best when the red and blue teams collaborate to strengthen the organization's security posture.

First and foremost, communication is crucial in this collaboration. To conduct exercises, there should always be communication between the various teams. Remember that the blue team's goal is to keep up with the latest technology and share that knowledge with the red team. This data will help enhance the organization's security. The red team must be informed of the most recent dangers and hacking tactics used by hackers and must advise the blue team about them. The purpose of an organization's test will decide whether the red team will notify the blue team about the upcoming test. If the purpose is to imitate a real-world scenario assault, they may not inform the blue team ahead of time, just to test their cyber defense mechanisms.

Management should encourage the teams to collaborate and communicate with one another. For the security program to continue to progress, improved coordination between both teams is required through effective resource sharing, reporting, and information exchange.

Cyber threat intelligence

Threat intelligence is a term that's often used by many professionals that encompass tactical, operational, and strategic intelligence. The sources, audiences, and forms of intelligence are all different. At the core, any threat intelligence that's received by the SOC, in any business, must be proactively actionable. The blue team should be able to absorb this intelligence and use it to proactively defend their organization.

In terms of the basics, threat data consists of indicators of various cyber threats such as IP addresses, URLs, or file hashes. These are referred to as **Indicators of Threats (IoTs)** or **Indicators of Compromise (IoCs)**. On the other hand, threat intelligence is a type of factual, processed, and provable record based on analysis that connects data and information from many sources to identify patterns and provide insights that would be relevant to the organization. It lets people and systems make educated decisions and take effective action to avoid breaches, fix vulnerabilities, improve the security posture of the enterprise, and decrease risk. Strategic intelligence usually focuses on the TTPs of the threat actors.

Often, such teams sit within the blue team. Alternatively, large organizations may prefer to have them separately and act as a standalone unit to collaborate across the blue team, red team, purple team, business lines, and more. We will discuss this in more depth later in this book.

Now that we have covered teams that work closely together with the blue team, let's understand the skills that organizations should look out for while recruiting. This will help ensure the right candidates are hired and placed in the right roles.

Skills required to be in a blue team

Blue teamers work with a pre-defined aim to secure the business network infrastructure and strengthen its cybersecurity posture. The methodologies and strategies they use to defend the network and systems from cyberattacks intertwine with each other. Management must have a better understanding of the goals and functions of the blue teamers.

Eager to learn and detail-oriented

To avoid leaving security vulnerabilities in a company's infrastructure, a very detail-oriented approach is required. Knowing how to create custom tools has several advantages. Writing software takes a great deal of practice and ongoing learning, thus the skill set gained aids any red team in executing the greatest offense strategies imaginable.

In-depth knowledge of networks and systems

A thorough understanding of computer systems, protocols, libraries, and well-known TTPs paves the way for the security personnel's success. A red team's ability to grasp all systems and keep up with technological advancements is critical. Knowing how to work with servers and databases will provide additional alternatives when it comes to discovering their flaws. Knowing how to use software packages that allow SOC analysts to monitor the network infrastructure for any unexpected or potentially hostile activities is very crucial. SIEM is a solution that analyzes security incidents in real time. It receives data from multiple sources and analyzes it according to a given set of criteria. Blue teams, similar to red and purple teams, utilize a variety of security technologies, including honeypots, sandboxes, XDRs and NGAVs, threat detection frameworks, and SIEM solutions. The following is a list of some of the most popular cybersecurity tools that are often used by these teams for their operational work:

- Splunk
- Haktrails
- Cuckoo Sandbox
- SecurityTrails API

Outside-the-box and innovative thinking

The cybersecurity team's major trait is their ability to think outside the box, always developing new tools and approaches to improve organizational security. To keep up with attackers, cybersecurity professionals must constantly think outside the box and uncover new tools and approaches. Cyber security teams deploy a variety of tools throughout their operations, including those for reconnaissance, privilege escalation, lateral movement, and exfiltration.

Ability to cross conventional barriers to perform tasks

SOC analysts always detect a good number of **False Positives** (**FPs**). To decrease the number of FPs they encounter on their SOC tools, sometimes, the senior SOC analysts have to cross several conventional barriers. They have to configure rules involving multiple filter criteria, which sometimes becomes overwhelming. Mind-mapping all the use cases helps these professionals as they would have to connect various use cases configured in the SOC tools. They would have to check whether certain rules that have been configured to serve a use case do not override other rules. Conflict resolution in the shortest time possible without the SLAs getting affected is very important. In many cases, this is like looking for a needle in a haystack.

Academics, qualifications, and certifications

Blue teaming roles do not require any kind of expensive certification or academic degree. Hands-on skills and talents are the most important for any blue teamer as this helps the professionals work better in any organization. However, having the right academic qualification and/or certifications may be considered good to have in various job descriptions. Many blue teamers are usually self taught and not spoon fed. However, some organizations may look for certain specific skills on the blue teamer's resume before shortlisting the candidate's profile for an interview. Hence, such academic accomplishments may end up becoming a shortlisting tactic, rather than a recruitment requirement by an organization. Some popular certifications in blue teaming are issued by bodies such as CompTIA, SANS, EC-Council, ISC2, ISACA, and others. There are multiple other technology/vendor-specific training programs and certifications that help blue teamers improve their hands-on skills on a given security product.

This section explained the skills needed and what talent to hire. However, this alone does not suffice. In the next section, we will cover talent development and retention.

Talent development and retention

One of the most challenging tasks of any security manager's life is finding a devoted, enthusiastic, and intellectual team member. It is a known fact that globally, there is a shortage of relevant skills. Hence, attracting the right talent to your organization is even more crucial. There is no single answer to this challenge. Let's look at a few ideas that management can implement.

Cyber labs

First, you can encourage employees to set up a home lab or use one provided by the company. Labs may be used to put real-world circumstances to the test, as well as to practice and master new abilities. For the vast majority of individuals, hands-on learning is the greatest way to learn, and in a lab, there is no chance of introducing risk into a production setting.

Capture-the-Flag and hackathons

Capture-the-Flag (CTF) competitions can be hosted at the company workplace. Such challenges help with cross-training, team building, and communication. CTFs and hackathons are the best staples for most of the young and vibrant cybersecurity conferences out there. They also offer any company one of the best places to locate fresh talent if they are trying to hire or expand the security team. Participants demonstrate not only their knowledge, but also their communication skills, teamwork abilities, and desire to assist and educate others.

Research and development projects

Developing an in-house project or finding some relevant projects from the open source communities is another possibility. Most open source projects require documentation or other help in various security areas. Security staff may find that this motivates them to showcase their skills in the public arena. So, for an organization to allow their staff to spend their time on such community projects could be seen as a magnet that attracts talent.

Community outreach

Allowing and encouraging staff to attend industry conventions or even local meetups is a great way to inculcate continuous learning habits. Attending a conference alone has its advantages, but the security staff may go further by preparing talks and presentations or even volunteering to help with the events. Moreover, this provides opportunities for the staff to network and build connections. This is a vital skill, especially for the **Cyber Threat Intelligence (CTI)** staff.

Mentoring

The company leadership team may help by mentoring young and fresh talent. Mentoring may be a great learning experience both on and off the work. This helps the security team learn more about the organization and feel more connected with the senior executives. Moreover, this motivates the staff to build a career path and network across the organization and business lines.

Continuous unhindered learning

The skills that are required to safeguard the business network are continually evolving as the cybersecurity industry adapts to manage emerging threats with new TTPs. Some studies showcase that in as little as 3 months, cyber professionals who do not continue to study fall behind and become much less successful. Tactics adopted by unethical hackers are evolving all the time; shouldn't the blue team staff evolve as well?

Helping staff continuously learn is critical for keeping the organization safe and secure in today's fast-paced cyberspace. Stakeholders are advised to adopt ongoing cyber training and reap the benefits of a high **Return on Investment (ROI)** in terms of security and productivity. Continuous and unhindered cybersecurity training allows the blue teamers to grow and refresh their knowledge while on the job, allowing them to keep current with industry trends. Even better, cyber professionals who have received on-the-job training perform the best to defend against attacks on time. Frequent training and certifications empower the blue team to swiftly detect and efficiently deal with incident response instances. Many firms invest in new, advanced security solutions to keep ahead of cyber threats. However, due to a lack of time or resources to understand how to utilize them, cyber professionals are sometimes unable to completely appreciate or apply the technology, resulting in them not having the edge over cyber criminals. To use new technologies, cyber experts must constantly learn new approaches and stay current.

Summary

It's not trivial to put an information security program together. Many programs are dysfunctional or non-existent, which contributes to the current state of business security. This chapter should have helped you understand the blue, red, purple, and CTI teams. An effective cybersecurity program requires organizational skills, knowledgeable, hardworking staff, strong leadership, and a very strong grasp of the cybersecurity niche.

In this chapter, we discussed the skills needed and what type of talent to recruit, and more importantly how to develop and retain that talent. In the next chapter, we will discuss how to manage such a team, as well as what indicators and metrics to set up to ensure the team is performing well and providing the organization with the most value.

2
Managing a Defense Security Team

In the previous chapter, we discussed the composition of a typical blue team, and how to hire the right talent. In this chapter, we will focus on how an organization's management team can ensure the blue team is working efficiently via measurable and tangible metrics that can be defined to ensure the organization is well defended.

Every organization should look at the right metrics that apply to them. This helps them not only objectively define the level of security they currently have but also ensure they are progressing and improving with each passing day. Moreover, in this chapter, we will review how to alleviate the workload on the blue team and look at automation with the help of some popular tools.

In this chapter, we will cover the following topics:

- Why must organizations consider metricizing cybersecurity?
- Why and how organizations can automate this process

Why must organizations consider metricizing cybersecurity?

Cyber veterans advise companies to measure the performance of the various cyber teams so that managers can manage their teams more effectively. Organizations have no idea how good or bad their **cybersecurity** posture is if they can't track ongoing security efforts. Cybersecurity is not something that can be completed once and then forgotten later. Cyber hazards are dynamically evolving, as are the techniques and technologies required to combat them. *Blue team* managers are advised to have procedures in place to evaluate the efficacy of the precautions that are deployed regularly.

Key Performance Indicators (**KPIs**), **Key Risk Indicators** (**KRIs**), and security postures give us a glimpse into how the blue team is doing over time. This assists the managers and the leadership in understanding what works and what does not, as well as making smarter decisions about future initiatives.

Metrics provide measurable and verifiable data that is frequently used to show management and board members that managers are working diligently to secure sensitive data and IT assets. Many **Chief Information Security Officers (CISOs)** and **Chief Information Officers (CIOs)** are discovering that reporting and giving context on cybersecurity metrics is a key aspect of their jobs, owing to the growing requirements from the shareholders, regulators, and board. Many board members in industries, such as finance, have a *fiduciary* or *statutory* obligation to manage cybersecurity risk and also protect PII data. Some new laws and regulations have helped fuel this trend. Some notable laws/regulations are **General Data Protection Regulation (GDPR)**, **California Consumer Privacy Act (CCPA)**, and **Lei Geral de Proteção de Dados Pessoais (LGPD)**, also referred to as the *General Data Protection Law* from Brazil, and many such others around the globe. In the next section, we will look at some of the main indicators that can be monitored by the blue teams to ensure compliance and security for their organizations.

Blue team KRIs

KRIs are measurements or metrics that a company uses to monitor present and prospective exposure to operational, financial, reputational, compliance, and strategic risks. The risks to an organization differ, depending on the workgroup or department. A retail bank branch, for example, might be concerned about bank accounts being created without adequate ID verification, while the financial institution's IT department will be more concerned with data security and breaches. An insurance claims department would be concerned with false claims KRIs, whereas an IT project management team would be concerned with server redundancy to assess and mitigate the risk of a system outage.

KRIs assess the risk associated with a specific action that the firm is considering, as well as the risk inherent in day-to-day operations. KRIs serve as a warning system for the firm, alerting them to financial concerns (*lost income*), operational issues (*lost productivity*), and reputational issues (*loss of credibility*). KRIs are used to evaluate the risk of possible adverse occurrences that may negatively influence a process, an activity, or an entire firm, which is commonly quantified in percentages. These metrics provide management with information about a company's technology and business risk profile, and they may be utilized to evaluate and enhance operations where improvements are needed.

There are several benefits an organization could gain from such an approach, as follows:

- They could give a heads-up so that proactive action can be taken, which leads to a lower number of security incidents and hence cost savings for an organization

- They could get a retrospective look at risk occurrences so that previous mistakes can be avoided

- They could inform the *board* or *senior management* if the risk appetite and tolerance have been met, or if any asset of the organization needs special attention

- The decision-makers and risk managers could have real-time actionable intelligence

Now that we understand what KRIs are, let's learn how a blue team should embark on the journey to develop them meaningfully.

How does a blue team initiate designing KRIs for their team?

Blue teams across industries usually follow a few phases to embark on the journey to understand which metrics work for them the best and how they would go about ensuring their organizations can get the most value out of them. It is important to note that these phases are iterative and must be revisited periodically. Over time, an organization's tech footprint or business needs may change, and the threat landscape will constantly evolve, leading to the KRI metrics having to be changed.

There are five main phases in this journey. Let's take a look.

Phase 1 – discovery

The discovery phase entails identifying all the IT assets across the entire organization, which helps cover the entire attack surface. Companies need to use this phase to find new, unknown, or unmanaged assets that are outside of their daily compliance processes. The blue team can then feed this initial information into more detailed asset scanning. This allows them to discover and begin to categorize the assets, based on the tech stack they are running.

This step involves discovering assets ranging from endpoints to servers, and network devices as well. This exercise should be done for both on-premise assets, as well as assets located on the cloud infrastructure. Moreover, discussions should be held with the respective business lines to understand the criticality of each asset. The blue team would then assign a risk rating to each asset, which is reflective of the business rating, as well as the threat rating of the asset. We will cover this aspect of risk rating in more detail in *Chapter 3, Risk Assessment*. The fundamental value of this step is to help the blue team identify the level and scale of security controls to be deployed to secure these assets.

Phase 2 – nominating the relevant assets and KRIs

After understanding the criticality of the assets, the next step would be to assign KRIs that are relevant, quantifiable, significant, and predictive to be implemented, to effectively monitor those assets. For successful risk management, it is advised to collect a suitable combination of leading and lagging indicators. Initially, blue teams configure a small number of KRIs that are challenging to track but critical for certain compliances or to track specific risks. Secondly, it is important to remember that all good metrics need to be tweaked to reflect the changing threat landscape and changing business strategies.

Phase 3 – baselining and limits

The next step is to define a baseline or an acceptable level of compliance for the asset for each KRI. Every baseline that's established during the initial phases of blue team development has to be revised periodically or deemed fit to keep up with the security posture. Hence, it is advised to determine and evaluate the thresholds that serve as triggers that violate established baselines, thus alerting the SOC team of non-compliance.

As an example, a server hosting a business application could be rated as high on criticality. Hence, the blue team may ask for 100% compliance on patching, antivirus, hardening, and other KRIs. However, an intranet portal to be used by internal staff may not be rated as high, and hence the baseline may not be as stringent. It is important to remember to define baselines that fit your organization's needs as well as the threat profile for the asset.

Phase 4 – monitoring, investigating, and reporting

Every KRI for the blue team is monitored regularly; the frequency differs based on the asset risk rating or maybe even the KRI in question. As and when certain non-compliances are flagged, the *security analysts* should immediately report and escalate them to the *SOC manager*, who, in turn, involves the respective teams/individuals to work on an action plan and mitigate the risk. Escalation management is usually done via the organization's ticketing system, which helps keep track of the open risks and the action owners assigned to each.

Note that **False Positives (FPs)** could occur here as well. Then, it is the responsibility of the blue team to investigate and close the ticket. There may even be a need to adjust the KRIs to ensure that the number of FPs stays low. Secondly, for a non-compliant asset, there could be a known business reason for not being able to meet the policy. In such cases, the blue team should work with the applicable teams to find other mitigating controls, or in some specific cases to accept the risk for a specific and acceptable amount of time.

Phase 5 – risk management

There is an adage that the cost of security cannot be more than the value of an asset. In real life, it is usually hard to quantify the absolute value of an asset, but in spirit, the rule of thumb is that the level of security must be adjusted based on the organization and the relative value of the asset. This is where risk assessments come into play.

We will cover risk assessments in *Chapter 3, Risk Assessment*, but it is important to note that KRIs are rarely ever a lift and shift approach. The blue team should understand the organization, the risks they face, and its tech stack before defining what works for them. They should also be open to regularly adjusting the organization's metrics:

Server Patching	
Compliant	60%
Partial Compliant	20%
Non Compliant	20%

Endpoint Patching	
Compliant	50%
Partial Compliant	15%
Non Compliant	35%

Network Device Patching	
Compliant	30%
Partial Compliant	0%
Non Compliant	70%

Server Hardening	
Compliant	80%
Non Compliant	20%

Endpoint Hardening	
Compliant	70%
Non Compliant	30%

Network Device Hardening	
Compliant	60%
Non Compliant	40%

User ID: Application Access	
Review Complete	95%
Review Pending	5%
Non Compliant	0%

Privilege Accounts: Servers	
Review Complete	92%
Review Pending	3%
Non Compliant	5%

Privilege Accounts: Desktops	
Review Complete	92%
Review Pending	0%
Non Compliant	8%

Antivirus Compliance	
Compliant	95%
Non Compliant	5%

Open Vulnerabilities	
Risk Open	1556
Risk Accepted	298

Open Audit Findings	
Risk Open	19
Risk Accepted	2

Database Patching	
Compliant	100%
Partial Compliant	0%
Non Compliant	0%

Database Hardening	
Compliant	97%
Non Compliant	0%
Review Pending	3%

Privilege Accounts: Databases	
Review Complete	20%
Review Pending	0%
Non Compliant	80%

Figure 2.1 – A sample dashboard

Selecting essential cybersecurity metrics

When it comes to cybersecurity KRIs, there are no hard and fast rules. These will be determined by your sector, the demands of your business, rules, standards, best practices, and, ultimately, your risk appetite. However, you should use metrics that can be understood by everyone, including non-technical stakeholders. If all your stakeholders do not comprehend them, you should choose different metrics or explain them better. Benchmarks and industry comparisons provide a simple way to grasp even the most complicated measurements.

Also, keep in mind that each KRI should sync up with an organization's policy or technical standard. Having a reliable policy is the foundation of KRIs. Such metrics should be seen as a way of meeting the compliance levels, as mandated by the organization. Hence, connecting each KRI with the relevant applicable policy helps give the right ammunition to the blue team to push for the requisite levels of compliance.

There are numerous recommendations around what areas and metrics an organization should typically focus on. **Center for Internet Security (CIS)** has created a list at `https://www.cisecurity.org/controls/cis-controls-list` of 18 controls that could be beneficial for a typical organization. This acts as a starting point and every organization should introspect and assess what is best for them. These controls are as follows:

- **Inventory and Control of Enterprise Assets**: Blue teams should be aware of the assets that belong to their organization. This includes servers, endpoints, and network components, located both on-premise and on cloud infrastructure. This acts as a starting point for the blue team to understand their network and ensure defense controls comprehensively cover all of the organization. Hence, having a KRI to measure the number of unknown assets is important.

- **Inventory and Control of Software Assets**: Similar to an asset inventory, there is value in the blue team being aware of a software asset register. This helps the defense team be aware of the tools and software in use within the organization and hence watch for any known vulnerabilities. Hence, having a KRI to measure the use of any unapproved or unauthorized software that's been identified on the network is of value.

- **Data Protection**: This KPI will help the blue team track the life cycle of the organization's data. There is value to ensure the data is identified, classified, handled, retained, and disposed of as per the security team's policies.

- **Secure Configuration of Enterprise Assets and Software**: As mentioned for the first two points, the recommendation was for the organization to track the list of their authorized assets and software. In this control, the organization needs to ensure each of these assets is deployed and configured securely. This KPI helps track an asset that does not meet the configuration standards of the security team.

- **Account Management**: This KPI tracks all the credentials that have been set up for any asset. This helps the blue team track if any credential was set up that deviates from the security processes. There is value in tracking privileged accounts (such as root and administrator) separately. This helps bring attention to these high-risk accounts.

- **Access Control Management**: This KPI tracks an organization's **Identity and Access Management** (**IAM**) processes. This should include KRIs around tracking any deviations from defined controls for access setup, revocation, and regular recertification. This helps ensure all the approved accounts are legitimate and still required by the organization.

- **Continuous Vulnerability Management**: Blue teams should run regular processes to look for any vulnerabilities across the entire organization and all assets identified in the first two controls mentioned. The frequency of these scans should be dependent on the risk rating of the asset in question. Tracking known vulnerabilities is an important KRI.

- **Audit Log Management**: Recording, collecting, and retaining audit logs from all assets is a very important control for security monitoring purposes. There are numerous legal and regulatory reasons to ensure logs are managed properly. Any asset that can't meet the log management standards of the organization should be tracked as a risk.

- **Email and Web Browser Protections**: An organization should consider securing all perimeter gateways, including email and internet proxy gateways. This will help the blue team defend their organizations and filter any threats coming in from these sources.

- **Malware Defenses**: Blue teams need controls to prevent and control the spread and execution of any malware. In case any asset becomes infected, there should be controls in place to quickly isolate and quarantine the asset to limit its exposure to the rest of the organization. Blue teams should track KRIs to ensure all assets have the requisite level of such controls.

- **Data Recovery**: Every organization needs to ensure every asset has regular data backups enabled on it. This helps in recovery efforts after an incident. An asset that deviates from the defined backup procedures should be treated as a risk and tracked with the help of KRIs.

- **Network Infrastructure Management**: There will be cases where employees bring their own devices to the office networks, or a guest or visitor tries to connect their device to your organization's corporate network. Such scenarios could be considered non-malicious, but there is also a possibility that an intruder tries to connect their machine to the network for nefarious purposes. A non-compliant asset may introduce threats to the entire organization, so there should be controls in place to isolate and then totally block such devices if needed.

- **Network Monitoring and Defense**: Every organization needs to defend itself against any attacks on its network. It's equally important to record logs and monitor them for any suspicious activities. The blue team should strategize the required defenses and track any deviations to ensure comprehensive coverage.

- **Security Awareness and Skills Training**: An organization's users could be the weakest link in the security chain. If your controls are underperforming, this statistic might provide extra context for evaluating their efficiency. Determining the percentage of stakeholders who have completed security training courses is one of the key indicators. Moreover, specialized sessions such as *Secure Code Development* will also be helpful for the relevant departments of the organization.

- **Service Provider Management**: An organization's threat landscape extends outside its physical boundaries, and the blue team's security performance metrics must reflect this. A formalized third-party risk management framework is a crucial KRI for the blue team. Similar to an organization's products, no two service providers should have the same level of criticality. Once again, a business-led risk assessment is crucial to decide on the level of scrutiny needed. A framework to manage and assess the security levels for all third parties will help manage this risk.

- **Application Software Security**: For all business applications, whether they are in-house developed or procured, should be regularly looked into for security threats or vulnerabilities. For in-house products, source code scanning processes should also be considered to review the application for security vulnerabilities. Typically, off-the-shelf products will have updates released regularly to patch any known vulnerabilities. These should be tracked by the blue team as well. Moreover, technology obsolescence should be monitored closely to ensure these do not bring any threats to the organization.

- **Incident Response Management**: Blue teams should be able to respond to any cybersecurity incident and recover from it. This requires careful planning and creating cyber playbooks. Regular testing of these plans is also essential to ensure their efficacy.

- **Penetration Testing**: Organizations should regularly test their defenses by mimicking the steps an attacker may take to breach their controls. These tests could be run by the blue team or trusted third parties. Some organizations may even run bug bounty programs, where the collective power of many security experts is tapped upon, to look for any vulnerabilities.

This section explained the typical KRIs that could be of value to an organization, and what blue teams should look out for. This may seem daunting. In the next section, we will look at ways to automate a lot of this work to help the blue team make better use of their time.

Why and how organizations can automate this process

Cyber risk management processes provide a comprehensive perspective of the cyber risks and threats a company faces. This enables authorized staff to assess risks and assign metrics to them, record changes in the organization's risk profile, and track risk and metrics against objectives and tolerance levels. Constructing a risk register is aided by corporate objectives and policies specified by top management, as well as other authoritative sources and standards. Risk assessment questionnaires are generated from the risk registry and used to conduct risk assessments. The results of risk assessments guide the creation and implementation of risk remediation or mitigation programs. Senior management is informed about these strategies, as well as the outcomes.

Risk management solutions can also be implemented for activities and responsibilities that are specified, quantifiable, relevant, and timely. Understanding the KRI standards and measurement criteria is critical to achieving this goal. Furthermore, numerous techniques and resources must be used to identify the organization's analytics suppliers and metrics consumers.

One of the most significant advantages of using technology to handle KRIs is that it eliminates time-consuming and inefficient manual labor. This helps automate data collection techniques and makes it simple to define acceptable thresholds and keep track of any issues and actions in the event of an incident. Moreover, such indicators can help meet regulatory, legal, and audit requirements.

What pitfalls to avoid when automating the workflows of the blue team

When automating the workflow of the entire blue team, you will face many hurdles, right from the initial phases of implementing the automation, such as a technical solution. The following pitfalls are encountered by many organizations and you are advised to watch out for to avoid them as early as possible while maturing your security program:

- **Lack of standards and best practices**: For better or for worse, standard tool operating methodologies look at exploring various features and capabilities that are used successfully by other organizations. Unfortunately, there are no standards across any sector and especially cross-sector that can be lifted and used as-is by a blue team.

- **Management orientation**: Senior management understands and expects a control that can explain the business benefits in clear dollar value. The notion of security metrics is still relatively new in the business. Many senior executives are unsure of its worth, let alone its design, and are unwilling to devote resources to its development.

- **The rate of change**: Technology evolves quickly, and risks that are incorporated or inherent in a specific technology today may rise or diminish when new versions or advances are released. Workflows related to a specific technology or even a technology-centric process should be re-evaluated regularly whenever the underlying technology undergoes a substantial upgrade.

- **Measures to maintain control**: Internal control measures must be established before effective indicators can be devised and executed. An organization that is unsure about its control measures will not be able to develop meaningful indicators around it. Fortunately, many organizations have gone through lengthy exercises to record essential control mechanisms as part of their compliance procedures. These controls are frequently used to determine active risk indicators on various security automation tools and can be used to develop measurable indicators.

- **Business risk management**: People in charge of establishing and managing the technology are primarily concerned with the technology itself, rather than the business risk posed by its failure. Moreover, the degree of impact a security incident could have on the business is difficult to measure. This creates friction between the blue teams and their leaders since they can't articulate the need for the controls.

Automating how KRIs are collected and presented

Once the blue team has worked on the KRIs, they need to define the requisite frequency of reporting each of them. This is where automation comes in and helps alleviate the workload on the team and focuses the attention of the team where it is needed the most.

KRI collection automation is enabled by a variety of use cases that have been built and fine-tuned using various security solutions already in place in the company's infrastructure. Various products and tools can help here. Some organizations use **Governance, Risk, and Compliance** (**GRC**) products to capture such metrics across various departments. On the other hand, **Security Information and Event Management** (**SIEM**) solutions can also be used to gather KRIs, measure and record them, and then present the data in the form of statistics on dashboards inside those security products. **Robotic Process Automation** (**RPA**) can also be used. RPA is where technology is used to execute specified tasks. Here, rule-based activities are performed to interact with existing applications.

The entire process of measuring, recording, and presenting KRIs may be automated. Many corporations have improved and matured their SOC for various compliance procedures because of automating KRI collection and visualization. This helps create a culture of data-driven decision-making and helps organizations focus on the risks that are most apt to them.

Dynamic dashboards may also be created to check if the KRIs have been set to the specific baselines and to monitor how often an asset goes out of compliance. Once these KRIs have been modified, alerts might be activated and delivered to the control owner. For some KRIs, exception reports or tickets may also be created as per the needs of an organization. Data science analytics may also be used to track different KRI dashboards, such as approved user creation and timely access revocation for exit and transfers. Before applying changes to KRIs throughout the network and systems, a continuous monitoring dashboard may be set up to ensure that all the changes are approved and appropriately confirmed.

As a result of a successful dashboard implementation, a blue team should feel empowered and be able to proactively protect their IT assets before they are hit with a cyber incident. At the end of the day, preventing incidents will always be cheaper and better than curing them.

Summary

In this chapter, we understood the importance of cybersecurity-related KRIs. We discussed how an organization should embark on this journey of setting up measurable and reliable indicators for themselves. We also spoke of automation and how some tools can be used to help the blue team operationalize this work and get the most out of this process.

Then, we looked at how the management team can keep a finger on the pulse of their organization to ensure its security is being measured effectively, as well as to ensure that the progress of the metrics is healthy and progressively improving as the threats and trends of the industry change.

In the next chapter, we will understand how risk assessments should be conducted in an organization and how they could be of benefit to the blue teams.

3
Risk Assessment

In this chapter, you will explore risk assessments, how they are done, why they are important, what values are calculated, and how they are calculated. You will learn how risk managers and other blue team members work in this field. The calculation of risk is primarily done before a threat occurs. It is a way that an organization can attacks, usually during a quarterly period. In this chapter, two examples will be used. The first involves a fictional hospital based in Frankfurt. The other example will be about a medieval castle, and we will be placing the important blue team figures in the landscape of the period.

In this chapter, we're going to cover the following main topics:

- Following the NIST methodology
- Asset inventory
- Risk management methods:
  - Threat identification
  - Risk calculation
- Risk management responsibilities

Following the NIST methodology

In this part, we will focus on the **National Institute of Standards and Technology** (**NIST**) methodology and how it shapes an organization. According to various experts, the NIST methodology is one of the most prominent methodologies used in the world today.

The **Department of Defense** (**DoD**) in the United States released version 1.0 of its NIST 800-171 Assessment methodology on November 7, 2019, because of a cyberattack on the DoD Navy submarine program in 2018, which caused a critical breach. Version 1.2 is the latest version at the time of writing and was released on June 10, 2020. Contractors first anticipated such a risk assessment methodology in January 2019, when Ellen Lord, Under Secretary of Defense for Acquisition and Sustainment, tasked the **Defense Contract Management Agency** (**DCMA**) with auditing the compliance of DoD contractors with the requirements of NIST 800-171. Of course, this was not the only reason why the DoD changed the NIST methodology. Attackers are becoming more prominent today and the world is changing, and information security methodologies need to change with it.

To generalize this methodology, there are six steps according to NIST involved in a **Risk Management Framework (RMF)**, which are as follows (as shown in *Figure 3.1*):

1. *Identify*: Essential activities to prepare an organization to manage security and privacy risks. Preparation can be the hardest topic regarding risk management, but once that is completed, a company is *ready, ready* to manage risk and achieve an RMF. The organization should identify and categorize the systems and information processed, stored, and transmitted, based on impact analysis. This impact analysis includes **asset inventory**, **risk calculation**, and **threat identification**.

2. *Select*: Select the set of NIST SP 800-53 controls to protect a system based on risk assessment(s). This is the most important step in an RMF, selecting what controls are needed for an enterprise. Due to complexity and impact, choosing controls is generally reserved for top management executives who have the experience required to build a system according to risk assessments, which can give an understanding of both where the organization stands financially and technically – does the organization have the right people regarding the tools that were chosen or not?

3. *Implement*: Implement controls and document how those controls are deployed. This can be accomplished by the risk management team, who use the systems we identified before and who can explain what each system does, and finally, how those systems should be secured. Documentation in this case is important so that if a control fails, we can find out effortlessly.

4. *Assess*: Assess to determine whether controls are in place, operating as intended, and producing the desired results. This can be completed by the risk management team, but it is preferable to include other teams in this procedure.

5. *Authorize*: A senior official, usually a **Chief Information Security Officer (CISO)**, makes a risk-based decision to authorize a system to operate. Unquestionably, the CISO has the final say, but if the team is not sure that the system will work, the CISO should be open to all suggestions and changes to the implementation before authorizing anything to go live.

6. *Monitor*: Continuously monitor control implementation and risks to a system by implementing risk assessments.

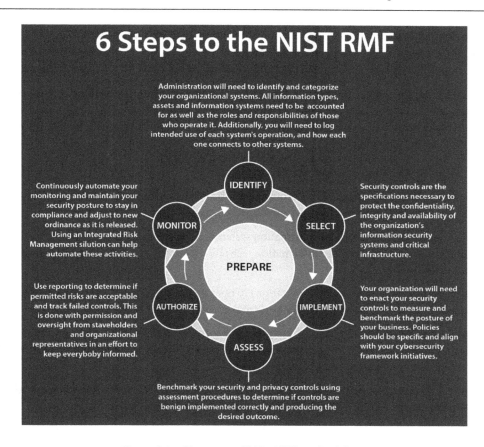

6 Steps to the NIST RMF

Administration will need to identify and categorize your organizational systems. All information types, assets and information systems need to be accounted for as well as the roles and responsibilities of those who operate it. Additionally, you will need to log intended use of each system's operation, and how each one connects to other systems.

Continuously automate your monitoring and maintain your security posture to stay in compliance and adjust to new ordinance as it is released. Using an Integrated Risk Management silution can help automate these activities.

Security controls are the specifications necessary to protect the confidentiality, integrity and availability of the organization's information security systems and critical infrastructure.

Use reporting to determine if permitted risks are acceptable and track failed controls. This is done with permission and oversight from staveholders and organizational representatives in an effort to keep everyboby informed.

Your organization will need to enact your security controls to measure and benchmark the posture of your business. Policies should be specific and align with your cybersecurity framework initiatives.

IDENTIFY

MONITOR   SELECT

PREPARE

AUTHORIZE   IMPLEMENT

ASSESS

Benchmark your security and privacy controls using assessment procedures to determine if controls are benign implemented correctly and producing the desired outcome.

Figure 3.1 – Six steps to RMF – NIST methodology

Completing these steps can be a rewarding venture. In the steps, a risk assessment is a requirement. Next, we will cover the NIST **risk assessment methodology**.

NIST risk assessment methodology

According to NIST, a risk assessment has the following straightforward steps:

- How to get ready for risk assessments
- How to manage risk assessments
- How to communicate risk assessment results to key personnel
- How to sustain risk assessments over time

Risk assessments are not one-time occurrences that provide permanent and conclusive information for stakeholders, guiding and informing responses to *information security risks*. Rather, they should be employed on an ongoing basis throughout any **Development Life Cycles (DLCs)** and across all tiers in an organization's risk management hierarchy. The frequency of those assessments and the resources should be measured against the expressly defined purpose and scope of the assessments.

Organizations conduct risk assessments to determine risks that are common to their core missions/business functions, common infrastructure/support services, or information systems. Risk assessments can support a wide variety of risk-based decisions and activities by organizational officials, including but not limited to the following:

- Development of an information security architecture

- An explanation of interrelationship requirements between any information systems (including systems supporting mission/business procedures and common infrastructure or support services)

- Designing security solutions for information systems and environments of operation, including a selection of security controls, information technology products, suppliers/supply chain, and contractors

- Authorization (or rejection of authorization) to operate information systems or to use security controls inherited by those systems (e.g. common controls used in Microsoft systesms)

- Modification of missions/business functions and/or mission/business processes permanently, or for a specific time frame (e.g., until a newly discovered threat or vulnerability is addressed, or until a compensating control is replaced)

- Implementation of security solutions (e.g., whether specific information technology products or configurations for those products meet established requirements)

- Operation and maintenance of security solutions (e.g., continuous monitoring approaches and applications, or ongoing authorizations)

As previously mentioned, always remember that risk assessment is bound by time. There will always be information systems, threats, and environments of operation, which will be modified over time.

In the next section, we will focus on *asset inventories* and how they are structured.

Asset inventory

In this section, we will focus on how a *blue team* member conducts an **asset inventory**. This takes place during the *identify* and *select* processes of the **NIST RMF methodology**.

An asset inventory should be in a table format. Let's take an example of a hospital organization that collects patient data and conducts medical services using that data. The primary goal of this organization is to help patients in Europe because this is a European organization.

The assets of this organization are its premises, hardware, devices, customers, and employees. Even though an asset inventory shouldn't be a **Human Resources (HR)** table, it's recommended that each system should be assigned to a *product owner*. A product owner is a person who handles the business behind each product/system.

When a blue team member writes an asset inventory, the first task is to give each asset a **Service Principle Name** (**SPN**). However, that should only be included in the asset inventory list, not the actual system. For example, a system's SPN is *HR-Operations*, but the actual service account name is *SRVASP-3258*, and this has nothing to do with its SPN, HR-Operations. Those identifiers can be used by hackers to identify which system they are attacking at every moment. But if we have many systems, and no indicator of which system should be targeted, then the hacker cannot target a specific system.

Having accomplished the first part of asset inventory, the blue team member now must identify where the system is, what its IP address is, and how that system works, which is a description of the system. Continuing with the aforementioned example, the hospital's asset inventory includes five systems. The first system's SPN is HR-Operations, its location is Frankfurt in Germany, and it's based on-site. Its IP address is `10.105.102.173/25`. The description of what this service does is as follows: "*The HR department of the hospital uses this service to catalog its employees and their personal data, including home address, educational history, and so on.*" The need to identify what type of system it is comes next. In our example, the system is Windows Server 2016, which is logged in to the internet and accessed by users using SSH.

Now, the blue team member must identify who the product owner is. It could be the HR *operations manager* or the *chief operations officer*; however, it is a best practice not to include c-suite members in the asset inventories as product owners, but this can only be done if we are talking about big organizations. Returning to our example, the product owner in our case is the HR operations manager, Michael Torstein. We should also be including an email address or other contact info for the product owner.

It's time to put this example into an actual table:

No.	SPN	IP address	Location	Description	System type	Product owner
1	HR-Operations	10.105.102.173/25	On-site, Frankfurt, Germany	The HR department of the hospital uses this service to catalog its employees and their personal data, including home address, educational history, and so on.	Windows Server 2016/internet access/the recommended use of SSH.	Michael Torstein, HR operations manager <Michael.Torstein @hospital.org>

Table 3.1 – An asset inventory (without controls)

This asset inventory table should be completed when we identify the assets. At the *select* phase of the asset inventory, we must place controls on the systems that we identified in the previous stage, thus adding to the table a **Controls/Protecting** column, where we add the controls assigned to each asset. Moreover, controls are assets, so they have to be added to the inventory along with which assets they are protecting. Another example we will use on this chapter is the example of a medieval castle. The king *Product Owner* has made a command to create an asset inventory of all his assets. He is aware that his castle has no walls to protect him but he is planning to change that soon. The blue knights have made their lists and assured the king that they will build the walls with the assistance of the townsfolk. Their goals is to protect the king and his people. Just like in a business, a product owner protects the products but he also has to protect the people making the products.

It's preferred to select the right controls for each asset. In every situation, a different type of control is preferred. If the goal is to protect a user's computer or an internal server, the protection mechanisms assigned to those assets are different.

For our example, to completely secure a server that was identified as an HR-Operations service, we need to get assurance that the service is only accessed internally so that we can include the relevant controls. Controls that should be included are a VPN service gateway, which protects the users who are accessing the server using SSH. Next, we need to include a *firewall* that can protect the service, along with an **Intrusion Prevention System (IPS)** and **Intrusion Detection System (IDS)** if the firewall doesn't include them.

Back to the medieval castle example: Having built the wall and set the ballistae on top of it, along with the wall guards watching, the need now arises to secure the inside of the keep, placing houses and guards who will do rounds inside. To do that on a business, there is an antimalware solution, which each system should include along with an internal firewall application.

No known brands will be used, as this is not a marketing book. Also, preferring a specific type of control will not protect a business if it is not helpful.

Let's place those controls in our asset inventory as before:

No.	SPN	IP address	Location	Description	System type	Product owner	Controls Protecting
1	HR-Operations	10.105.102.173/25	On-site, Frankfurt, Germany	The HR department of the hospital uses this service to catalog its employees and their personal data, including home address, educational history, and so on.	Windows Server 2016/ internet access/the recommended use of SSH.	Michael Torstein, HR operations manager <Michael.Torstein @hospital.org>	**On the perimeter**: A firewall, an IPS solution. **Included on the server**: An internal application firewall, and an antimalware solution.

Table 3.2 – An asset inventory with controls

Table 3.2 shows the way the asset inventory should be worked on after the controls have been selected. In the next section, there will be guidance on threat identification, which is the next part of the NIST guidelines to assess risks.

Risk management methods

In this section of the book, we will be building a methodology for risk management. This methodology is not the only way that a RMF can be used but it is one of the methods that are out there.

Threat identification

In this section, we will consider the fundamentals of **threat identification** and how that is accomplished by the blue team.

In order to identify what threats we have to deal with, we must first identify what risk model is used in an organization.

- A **threat** is any situation or event with the capability to destructively affect organizational processes and assets, individuals, and other organizations, through an **Information Technology (IT)** system via unauthorized entry, destruction, disclosure, alteration of information, and/or denial of service. **Threat events** are caused by **threat sources**. A threat source is categorized as follows:

- The purpose and technique targeted at the abuse of a vulnerability

- A circumstance and method that may accidentally exploit a vulnerability

In general, types of threat sources include the following:

- Hostile external cyber or physical attacks

- Internal human errors of neglect or delegation
- Structural failures of organization-controlled resources (e.g., hardware, software, and environmental controls)
- Natural and man-made disasters, accidents, and failures beyond the control of an organization

Some classifications of threat sources use the type of hostile impact as an organizing principle. Multiple threat sources can initiate or cause the same threat event – for example, a provisioning server can be taken offline by a denial-of-service attack, a deliberate act by a malicious system administrator, an administrative error, a hardware fault, or a power failure.

Having understood what a threat is, we now have to define what a risk model is.

A **risk model** is defined as risk factors that can be assessed and the dependencies among those factors. **Risk factors** are features used in risk models as inputs to determine the levels of risk in risk assessments. Risk factors are also used extensively in risk communications with the board or the rest of the security team to highlight what strongly affects the levels of risk in particular situations, circumstances, or contexts. Distinctive risk factors include threat, vulnerability, impact, likelihood, and predisposing conditions. Risk factors can be broken down into more detailed characteristics (e.g., threats broken down into threat sources and threat events). These definitions are important for organizations to document prior to conducting risk assessments because the assessment will rely upon well-defined metrics of threats, vulnerabilities, impact, and other risk factors to effectively determine risk.

Returning to our previous example, a hospital can be considered a high-risk area because, in hospitals, people get treated and usually we don't know who will be treated beforehand. Therefore, having a server that includes personal data inside a hospital can be at higher risk of being accessed by outsiders than it being somewhere other than the hospital, but it is still accessible through a network using a VPN.

Now that we've established where the *treasury* is going to be, we need to ensure that the location remains confidential and only known by the product owner (also known as *the king*), and any administrators, also known as (*treasurers*), that need to go on-site and fix any issues that might occur. Access to this system should be done via a VPN, also known as the *secret tunnel*. Managing that confidential approach should be the **Chief Information Security Officer's (CISO's)** job, who is also king for any of the controls that the blue team is using.

Identifying threats is one of the most important things to do before building a risk assessment. To accomplish that, it is recommended to use risk factors. In those risk factors, identifying vulnerabilities of threats that may happen is the first step. If our risk model allows some threats through, we must understand that there can't be a 100% secure system, but the risk model should define those exceptions.

For example, if one of the administrators is allowed to have administrative access to the HR-Operations system, then the risk level of that asset will be higher because there is a hole in that system. If that user is not trained properly in information security awareness, then the system is condemned to be breached somehow.

What will the next threat to the treasury be? After we've determined the risk value of the hospital as a location and have placed the server securely in an off-site location, we next need to see that any attacks on the treasury are kept at bay by the controls. One of our treasurers has a friend who is a fugitive from the law of another country or state. The treasurer decides to let the fugitive through the secret tunnel either accidentally or is persuaded through manipulation.

This is a well-known threat that security professionals call *spear phishing*. In this example, the administrator willfully or unconsciously let the attackers in through the gates and into the secret tunnel. The likelihood of that happening can be determined by calculating what training the administrator got before the incident and whether the administrator reported the incident to the king, before letting the fugitive inside. The impact of this could be high; however, the impact could be minimized if we used least-privileged controls – for example, by not letting the administrator have too many privileges on his machine.

According to the predisposing conditions that were identified, one of them being a super-user to the HR-Operations service, we need to put those factors on our asset inventory next to the controls we placed before. There should also be any vulnerabilities we identified for a particular system or version of the control/system we have placed. There also has to be a threat column, where we define the threats – in our case, spear-phishing, social engineering, and malware. The threats should be placed on all columns because they'd be the same across all assets.

No.	SPN	IP address	Location	Description	System type	Product owner	Controls protecting	Threats
1	HR operations	10.105.102.173/25	On-site, Frankfurt, Germany	The HR department of the hospital uses this service to catalog its employees and their personal data.	Windows Server 2016/internet access/the recommended use of SSH	Michael Torstein, HR operations manager <Michael.Torstein@hospital.org>	**On the perimeter:** A firewall, an IPS solution. **Included on the server:** An internal application firewall, and an antimalware solution.	Spear-phishing and malware

Table 3.3 – An asset inventory with identified threats

In this next section, we will tackle *risk calculation*, which is the next step in the risk assessment process.

Risk calculation

Calculating risk is a messy situation; however, it can become a rewarding experience once you learn how to do it. A blue team member should be aware of the level of **security awareness** in an organization in order to calculate the risk.

In our scenario, the super-user has gained enough security awareness to be in a medium-risk position, and that means that although their work is valuable to an organization and their contribution is valued highly among their colleagues, the risk level of their accessibility is still questionable.

As can be gleaned from this example, the likelihood of the spear-phishing threat could be a medium or low risk; however, a good risk manager knows that the likelihood of occurrence is a weighted risk factor, based on an analysis of the probability that a given threat is capable of exploiting a given vulnerability (or set of vulnerabilities).

The likelihood of occurrence is a risk factor that combines an estimation of the likelihood that the threat event will take place with an estimation of the likelihood of impact. For adversarial threats, an assessment of the likelihood of occurrence is typically based on the following:

- Adversary intent
- Adversary capability
- Adversary targeting

Other than adversarial threat events, the likelihood of occurrence is estimated using historical evidence, observed data, or other aspects. Note that the likelihood that a threat event will take place or happen is evaluated in relation to a specific time frame. If a threat event occurs in that time frame, the risk assessment may take into consideration the estimated frequency of the event and how many times the spear-phishing threat occurred.

The likelihood of threat occurrence can also be based on the state of an organization (including its core processes, enterprise architecture, information security architecture, information systems, and environments in which those systems function) – taking into account predisposing conditions and the presence and effectiveness of deployed security controls to protect against unauthorized behavior, detect and limit harm, and/or maintain business capabilities. The likelihood of impact addresses the probability (or possibility) that a threat event will result in an adverse impact, regardless of the measure of harm that can be expected.

Having seen the word *impact* mentioned a lot at this stage, it also must be calculable. The level of impact from a threat event is the magnitude of harm that can be expected to result from the consequences of unauthorized disclosure of information, unauthorized modification of information, unauthorized destruction of information, or loss of information or information system availability. Such harm can affect a variety of stakeholders, including, in essence, anyone with a vested interest in an organization's operations, assets, or individuals, including other organizations in partnership with the organization:

- The process that is used to conduct impact determinations
- Assumptions related to impact determinations
- Sources and methods for obtaining impact information
- The rationale for conclusions reached regarding impact determinations

Of course, all that can't be fitted in an asset inventory, but the conclusion of how much impact there'd be if a threat became reality should be enough.

In the hospital example, if the threat of spear-phishing becomes reality and the super-user who is security-aware still fails to recognize the phishing attempt, meaning they get fooled by the phishing attempt, then the impact that this will have on an organization is high. Usually, the impact is calculated financially, meaning how much money the organization would lose if that asset became unavailable or inoperable. Those calculations should be assessed by the **Chief Financial Officer** (**CFO**) of the company.

In order to calculate the risk, we multiply the likelihood by the impact. In our example, the impact is high because this is the HR-Operations service, and it is invaluable to the operation of the hospital. The threats are spear-phishing, social engineering, and malware. The threat of spear-phishing and social engineering has a medium likelihood, which means that the risk of the organization getting attacked with this threat is high. However, the malware threat has a low likelihood, so the risk level is medium. The CFO can quantify that risk by calculating how much the asset costs and how much each bit of data costs. This technique can help organizations calculate risk using a financial report.

It's time to place those calculations on our asset inventory board. Remember, the impact will be calculated based on a low, medium, or high value. We will also be calculating data by saying that each **MB** of data has a $2 value to an organization. If the organization holds 10 **GB** of data for all employees, then the initial value of the asset is $20,480. Then, after that most information security professionals would use this calculation along with the value of the server, which could be $1,520. Then, we can summarize that this calculation would lead to an initial value of $22,000, and then the blue team member would multiply this amount according to the risk level of the threat, which could be 1 for low, 2 for medium, and 3 for high. The threats were calculated as social engineering and spear phishing – that is, high – which makes the final estimate $66,000. For the medium threat of malware, the final estimate is $44,000. This is not an actual representation of how a hospital would calculate its data, but you can understand the proportions this could reach.

No.	SPN	IP address	Location	Description	System type	Product owner	Solutions protecting	Threats	Likelihood	Impact	Risk level
1	HR operations	10.105.102.173/25	On-site, Frankfurt, Germany	The HR department of the hospital uses this service to catalog its employees and their personal data.	Windows Server 2016/ internet access/the recommended use of SSH	Michael Torstein, HR operations manager <Michael.Torstein@hospital.org>	**On the perimeter:** A firewall, an IPS solution. **Included on the server:** An internal application firewall, and an antimalware solution.	Spear-phishing and malware	Spear-phishing = medium, and malware = low	High – the initial value of the asset: $22000	**Spear-phishing = high ($66,000)** **Malware = medium ($44,000)**

Table 3.4 – An asset inventory with risk, likelihood, and impact

In the next section, we will be looking at the responsibilities behind the risk assessment process, who is responsible for the risk management framework, and what their specific role should be.

Risk management responsibilities

During this section, we are going to cover the people responsible for risk management in organizations; however, we should always remember that everyone in an organization should be responsible for the risk they pose to it by becoming security-aware and not being idle to threats that the organization faces daily.

Regarding the people responsible for risk management, first and foremost is the CISO, who is in charge of risk mitigations by providing solutions to problems the organization will face or is facing at any moment. Furthermore, the CISO has to be in control and protect the organization from all threats that are stacked against it. The CISO chooses to identify the risk model and what analytical approaches should be used in risk assessment. Also, the CISO will be the one to communicate the risk assessment results to the board along with the solutions that will minimize that risk. Some organizations may not have a CISO position, but this can be filled by an IT manager/director if they understand the need for risk management and can take on this crucial position.

The CISO needs people to work under them. One of the most important people to work with the CISO and who is invaluable to the risk management process is the **Security Awareness Officer** (**SAO**), who will take care of educating employees. An SAO is responsible for calculating the educational level of the employees and whether they have completed all the training assigned to them. One of the SAO's additional responsibilities is to test people who've gone through the training by orchestrating phishing campaigns, which should include subjects relevant to an organization. For example, for the hospital, it could involve changes to regulations, which the HR manager would be curious about. Anything causing curiosity should be considered for a phishing campaign.

The next officer who works with the CISO is the **Chief Risk Officer** (**CRO**), who is responsible for calculating the risks that an organization is facing. The CRO is also responsible for identifying threats and making sure that all parts of the organization, especially the SAO who is responsible for educating the employees, understand the threats that are out there and that they are addressed according to each employee's role. The CRO will communicate with the CISO about the results of a risk assessment and other risk-related information so that the CISO can communicate with the board and take the measures necessary.

In conclusion, we've seen the need for a risk management attitude that needs to exist in all businesses, not just cybersecurity ones. It is time to build a world that knows about threats and their consequences, as well as the risk that accompanies them.

Summary

In this chapter, you have seen how risk assessments are completed, with examples such as the hospital and the castle showing how risk assessments were completed, even in times of yore. You have learned about risk management frameworks and how they are implemented in organizations using the NIST methodology. Also, you have seen the proportions the threats can reach for an organization when financial reporting.

References

- *NIST SP 800-171 DoD Assessment Methodology* (cuick trac) (`https://www.cuicktrac.com/nist-compliance/nist-sp-800-171-dod-assessment-methodology`)

- *NIST Risk Management Framework Webcast: A Flexible Methodology to Manage Information Security and Privacy Risk* (NIST) (`https://www.nist.gov/news-events/events/2019/02/nist-risk-management-framework-webcast-flexible-methodology-manage`)

- *Protecting Controlled Unclassified Information in Nonfederal Systems and Organizations* (`nist.gov`) (`https://nvlpubs.nist.gov/nistpubs/SpecialPublications/NIST.SP.800-171r2.pdf`)

- *What is a Risk Management Framework?* (`https://www.cybersaint.io/glossary/whst-is-a-risk-management-framework`)

- *Guide for Conducting Risk Assessments* (`https://nvlpubs.nist.gov/nistpubs/Legacy/SP/nistspecialpublication800-30r1.pdf`)

4

Blue Team Operations

From what we've seen thus far, we can understand that a blue team requires dedication along with loads of documentation, which is a must if the team wants to achieve great breakthroughs. In this chapter, we will explore **blue team operations**, the different areas where those operations take place, and how they are completed. You will learn about how **operation managers** and other blue team members perform their operations in different fields.

In this chapter, we are going to cover the following topics:

- Understanding defense strategy

- Blue team operations – infrastructure

- Blue team operations – applications

- Blue team operations – systems

- Blue team operations – endpoints

- Blue team operations – cloud

- Defense planning against insiders

- Responsibilities in blue team operations

Understanding defense strategy

Defining and implementing strategy is a **Chief Information Security Officer's (CISO)** responsibility when they enter an organization. What that means is that someone must predict a company's impact regarding information security and that will be the CISO. When all risk assessments are completed without any controls in place yet, that is when the time comes for a **defense strategy**.

Defense strategy includes the controls and procedures required to secure an organization. In this chapter, we will learn how to build strategic procedures and place controls in an organization's perimeter and within the organization itself. Strategic thinking must be followed by strategic action; each time we deem something as important, we have to build toward that goal. This means if an organization buys all the equipment and technology it may need without calculating and prioritizing the need for it, then the organization is doomed to fail. Many entrepreneurs before understanding a threat and its significance buy tools, without calculating the need for those tools first. Remember, it is not obligatory that every business has every type of control, so let's see that in an example.

Let's recall the example mentioned in the previous chapter, with the castle and its king. The king will not wait for a war to break out in order to build the walls and buy the ballistae. He will buy it and set up those defenses before the attack happens. Some business executives may not understand this concept, unfortunately, because start-ups nowadays tend to wait for a bad thing to happen, mitigate the attack, and then buy the solution so that it doesn't happen again. But in our example, the king would be ousted before the ballistae arrives if he waited for the attack to happen. Probably, if the king hired some mercenaries, he would survive, but still the attack would be catastrophic if he didn't have defenses.

Thinking strategically should be every stakeholder's goal, not just the CISO's. There needs to be trust among the stakeholders. Trust plays a really important role in all business. Most businesses call this commitment to a company's goals, but generally, it has to do with trusting the people you work with. If there is no mutual trust and employees don't trust one another, there will be chaos. Thinking about adopting a zero-trust principle, this could cause dysfunction in small teams, but in bigger teams where you know who is protecting you, zero-trust can help prevent issues such as a stakeholder having too many privileges and making a mistake that might cost the company money or reputation. The blue team is able to mitigate those attacks before they become reality and cause catastrophic failures, protecting the king's castle.

The blue team should be preventative, but it can also mitigate issues happening around us every day. For example, if the king's blue knights defend the king's chamber from the assassins that came one night through the castle's walls to kill the king, then that is the definition of mitigation. However, the strategy to place them there was made by the CISO, who predicted that the attack would be coming in that specific period and location.

The CISO had to calculate the timeframe so that the knights would be stationed close to the king in order to stop the assassins.

Let's use another example, with the hospital featured in the previous chapter. We are in the phase before controls are placed on the HR operations server. The CISO needs to calculate the need for controls and explain why he needs them to be there, what those controls will do, and how will they protect the organization. The CISO creates a mind map where he places the external and internal controls with a need and a timeframe.

In our example, the need for a firewall is high. As an external control, the firewall is needed, so we must make sure it exists within the organization on the first day that the asset is placed on the external premises.

Next, we need a way to access that server. The use of SSH is good but not perfect, so the CISO places another control, a *VPN connection*. This control would also be needed on that first day.

With this control in place, the CISO now must monitor the system, either by placing controls inside it or making sure that the server has internal controls. The CISO chooses to recommend an antimalware internal system, an internal firewall, and an external IDS/IPS solution if that was not included in the external firewall. Those solutions can be placed on the first day; however, that decision would be left to the provider of those solutions.

When a spear-phishing attack happens, the people working on the server notice that something doesn't look right and inform the blue team to mitigate the issue. Even if sometimes spear-phishing attacks can evade the IDS/IPS solution, meaning that an email gets through to the server, the reporters, who are the employees of the organization, must have the security awareness needed to detect the issue before it becomes an incident and report that to the blue team.

Therefore, the CISO also must calculate the training required for all employees that use that server. The training can include phishing simulations, phishing campaigns, and so on. A phishing simulation is a presentation where the CISO has three emails and asks the people attending the presentation to differentiate between a real email and a phishing email. After that is done, the CISO presents which one was the real email or the phishing email and explains why and how to detect those emails. This training technique is good for small organizations; however, for bigger organizations, there are phishing campaigns. A phishing campaign is a method that sends an email to all the people in an organization and has a link that they must not click. If the organization has a reporting mechanism in place, the end user can report that email to the blue team and then get approval – for example, through a happy emoji or a rating system – that the user detected the campaign and doesn't require extra training. A phishing campaign is a great way to identify users who need additional training to detect phishing emails.

In the next part, we will focus on the infrastructure of the blue team and see how it works.

Blue team operations – infrastructure

The blue team doesn't exclusively mitigate attacks, but they strengthen an entire digital infrastructure by performing vulnerability assessments and making sure that other teams can work without interruption from outside forces.

Let's return to our castle. The blue knights are now assigned to patching the walls of the castle. A menial task, to be sure, but it is still a requirement for any future attacks. They also check that the ballistae are working properly. Their assignments include checking the guard shifts and seeing whether any of them are lazy or not doing their work as intended.

Generally, a blue team operates in an organization by checking that everything works as it should. Of course, there is a collaboration with different teams to ensure that, such as collaboration with the red team on bypassing the firewalls and the IDS/IPS system or breaking into the server.

Returning to the hospital example, the blue team needs to limit the attack surface of the HR operations server, meaning lowering the number of users that can use the server or other methods, such as having a principle of least-privileged users. Afterward, they need to mitigate any poor design flaws that might come up while the server is active by applying network segmentation or other methods. Next, the blue team will have to ensure that there is log monitoring enabled on the server. By enabling log monitoring, the team can track any changes to files within the server to ensure accountability, meaning which user did what and for what reason. This can also prevent or mitigate malicious insider attacks. The blue team also needs to ensure that the network traffic to and from the server is encrypted, which is why the CISO applied the VPN measure.

This way, the blue team can identify any vulnerabilities and the level of ability the hacker would need to gain access by using its own **red team penetration testing**. When those two values have been identified, the blue team will try to patch the vulnerabilities, thus raising the level of the ability the hacker would need to gain access. At this point, the red team is able to detect any zero-day vulnerabilities and report them to the vendor of the infrastructure in order to create updates that the company can take advantage of.

Finally, security audits – either internal or external – are a great measure to prevent infrastructure attacks. The blue team can use those audits by applying the measures recommended by the auditors or making sure they follow compliance measures for different infrastructures.

Having made clear the ways that the blue team applies measures to protect infrastructure, we will now be looking into application security and how the blue team plays a critical role in that.

Blue team operations – applications

Applications, like infrastructure, require measures so that the blue team can apply for better information security during the development and production phases. Applications can also be vendor-related, but we already covered that in the previous section.

Placing measures before development can include creating an **information security championship**, where developers can gain knowledge and have someone who knows the security landscape as a champion, who can keep track of any discrepancies that the development team might not notice. A champion must stay aware of the affairs of the team and prevent any outside forces to disrupt the team's activities. This can also be the role of a scrum master; however, an information security champion focuses on any vulnerabilities that might occur during development and has to make sure that they are covered before the product release. A champion is still a developer, so their responsibility is to fix the vulnerabilities the blue team uncovers. If they can't, then the blue team reports this to any vendors that might be able to cover them, by applying updates to the software the team is currently coding with.

This method is better applied before developers start coding. During development, a good method that a blue team cab apply is vulnerability assessments on the parts of the software that are complete. Agile methodology calls those parts **iterations**; if each iteration is scanned for vulnerabilities, then the software will be able to avoid any vulnerabilities that might come up during development.

Because the vulnerability landscape changes drastically every single day, after the development of the product is complete, the blue team has to do vulnerability assessments at regular intervals and submit bugs to the developers, who have to make sure those bugs are covered. In this case, a great methodology to use is **kanban**. Through kanban, the blue team can ensure that the development team has a kanban board that they can use to report any bug. The blue team can fill the board with work items so that the team can choose which one to work on at a specific time and where to place it in the work in progress of the Kanban board. After the work in progress is complete, the team can evaluate again whether the vulnerability is covered by scanning for the vulnerability again. If the vulnerability is not found, then the work item can move to a **done** or **complete** column. The blue team must remember that vulnerability assessments are a continuous process, not a one-time deal. Covering a vulnerability doesn't mean the threat is over; it might even open the software to new threats that need to be addressed before any incident occurs. Sometimes, software or programming library updates introduce new vulnerabilities that need to be constantly monitored, both during the development and production phases of the software life cycle.

There are also many best practices that should be followed during and after development. Because those practices are also changeable, teams need to ensure that they have included the best practices during and after development.

Making this simple, with our castle example, the blue knights want to create a procedure for the guard shifts inside and outside the perimeter of the castle. They place three guards around the king's chamber – two guards at each door and one guard that patrols the perimeter of the chamber. Having protected adequately the king's chamber, now the blue knights place guards at the main gate of the castle. Three guards are placed there again. Next, guards have to be placed on every corner of the interior of the castle. All guards are equipped with torches to see whether anyone tries to break into the castle. Developing guard posts would be a good measure to let the guards take a breather from time to time. After they have set up everything, the blue knights perform regular inspections of their procedure to see what is going well and what is not and make relevant changes.

Returning to our example with the hospital, the HR operations team needs to create an application that controls all data of employees by allowing the employees themselves to change any data that they might want to. For example, if an employee wants to study a course at a university, then they can use this system to add new educational data to their profile.

The developers are ready to start the development. They have selected amongst themselves who will be the **InfoSec champion** and the **scrum master**.

During its first **sprint planning**, the development team must understand the needs of the mission, set their own goals in accordance to that, set a code of conduct between them, and allow the product owner to create a product backlog, which includes a list of work to be done during each 2-week period, which is a sprint. After the first 2 weeks, the team will have done 14 daily scrums for each team member. Daily scrums are short talks, which can be as short as 15 minutes each, among the developers that help them better understand any discrepancies and challenges that might have occurred during each day of development. After these are done, the team assembles for a retrospective on the 15th day, where during that time it is advisable to include the blue team members taking care of the vulnerability assessments. During each retrospective, if the blue team can prepare a list of vulnerabilities either on the spot or the day before each retrospective, then the team can discuss during the sprint review what should be added to the backlog, including the vulnerabilities identified. During this time, it is good to review the best practices that were applied during each sprint.

After the team has finished as many sprints as required for the final product and it is online and working, the development team now changes its tactics to kanban. During this time, the development team receives the blue team's work items to fill the kanban board. Then, the team picks up each bug and fixes it, trying not to disturb operations, and performs updates on any day that is less likely to disturb the operation of the system. Any updates from best practices should also be included after the development is finished.

In the next section, we will see how the blue team handles systems, what those systems are, and how they are taken care of in blue team operations.

Blue team operations – systems

When talking about systems, we are not just referring to computers but also other types of information technology, such as scanners or other mobile technologies.

A system is considered a generic supporting methodology or technology used by businesses that, if implemented, can lead to any results required by the business. It can be anything. Blue teams have to ensure that those systems are safe and protected from any alterations to their normal usage. Those alterations could be changes to camera footage, deletion of camera footage, changes in code for scanners, and so on.

If those situations happen, the business will likely lose revenue, and disasters and incidents will occur, which can be avoided if the blue team remains vigilant and creates measures that will not allow any external users to access the systems and ensure internal users are really careful when using any of the systems.

Returning to our castle, the king has established a post office in his domain. The post office uses carrier pigeons and messengers who carry mail to different lands. The king has sent his red knights to attack another castle. He tries to use the messenger birds first, but he doesn't know whether his knights received the message. Therefore, he sends a messenger with an escort of blue knights accompanying him as a measure of protection. The messenger gets attacked by bandits on the border; luckily, the measures taken before ensured the system works, so the messenger survived and delivered the news to the red knights. Using the same system, the king sends reinforcements and wins the battle of the enemy stronghold.

For our hospital example, the blue team follows the NIST methodology and creates three systems for storing patients' data. One system will be used by the doctors in the hospital, where they will have access to the patient database by logging in to the VPN and afterward to the database server, using queries to find each patient's data. The other system will be used by the nurses, who will have to use the same security measures in order to get access. The third will be used by the financial department and will include financial and insurance data of the patients, and this will hold the same type of security measures. Using the principle of least privilege, the blue team will allow each employee to have access to the data they need, not all the database data, as that will cause problems for the business if any of that data is leaked.

Using all the recommended methods of securing any system can be cumbersome, both financially and ergonomically, but working in information security teaches you how to calculate a need by dividing the resources you have at your disposal and, thus, fulfilling that need as much as possible. Afterward, you accept any risks that could not be covered so that you can cover them in the next release. Last but not least, always cover the critical vulnerabilities of any system first before anything else.

In the next part, we will focus on endpoints and how endpoint security is the reason we need the rest of the information security landscape, in order to protect our employees from outside interferences.

Blue team operations – endpoints

The real point of endpoint security is to prevent hackers from accessing any systems that endpoint users might have access to. In order to secure an endpoint, we have to place the right controls internally and externally with the same methods we used for the infrastructure and systems. The blue team takes care of those endpoints, making sure that they can monitor those systems. Of course, installing those systems should be the work of the service desk team, but if the team is not big enough, the blue team can handle the installations as well.

In the COVID-19 era, we've seen the rise of remote working conditions. Remote systems are harder to control and regulate properly. The blue team cannot ensure that the endpoints are in a safe location, so they must ensure that the internal controls are all in place to protect end users.

Generally, if an end user is using a public network, then the risk is higher for threats to become reality than when using their home network. In this case, the blue team has to enable certain controls, such as a VPN and the principle of least privilege, on the machine they want to protect.

Let's take our castle example. As we discussed in the previous chapter, the king created secret tunnels for his servants to come and go from their homes, which is the VPN measure, and the tunnels are kept tightly shut. The king ordered guards to be placed outside each of his servants' houses. This could be an antimalware system. Then, the servants locked their homes so nobody unwanted can enter them. They have also bolted the windows. This could be a login screen measure. Their homes are also built with stone. This could be the internal firewall measure.

Let's take our hospital example. During the COVID-19 pandemic, all the employees of the hospital moved to their homes and worked remotely. This made it difficult for blue teams to do their work to protect the end users. Productively, the team predicted this problem, so they installed internal firewalls, VPN applications, network controls, and antimalware solutions and enabled the principle of least privilege, thus not allowing users to install any dangerous software on their business laptops. They also created a safe browsing space where tend users were not allowed certain websites, thus stopping future attacks and assuring a safe experience for the end users working from home.

Commonly, when endpoints have moved outside an organization's premises, they tend to be harder to regulate, monitor and control; however there are systems that can do those three things – for example, **Remote Access Control (RAC)** for different purposes, such as installing the software needed for the business, fixing computer issues remotely, and doing service desk-related procedures without a physical presence in the home-working area.

In the next part, we will talk more about cloud security and how that works for a blue team.

Blue team operations – cloud

Cloud security is an area where blue teams can always receive help from other teams, usually related to the cloud vendor. There are three different types of services, **Infrastructure-as-a-Service (IaaS)**, **Platforms-as-a-service (PaaS)**, and **Software-as-a-Service (SaaS)**. The way this works is that a blue team can be given areas that they have to cover. No type of service is the same and there can't be two types for one infrastructure. However, there are some similarities between the different types. Of those similarities, the first one is the abstraction, pooling, and sharing of scalable computer resources across a network. Every type of service also allows cloud computing, which is the act of running workloads within that system. Using a unique mix of technologies, every cloud is almost always created with an **Operating System (OS)**, a management platform, and **Application Programming Interfaces (APIs)**. Virtualization and automation software can also be added to increase the efficiency and capabilities of every kind of cloud service.

If the cloud service is an IaaS, then the cloud security landscape for the blue team is bigger because the vendor only deals with the infrastructure, which is loaned to the organization and includes the servers, storage, virtualization, and networking, all of which are managed by the vendor. The organization has to manage the applications, data, runtime, middleware, and OS. Thus, in this case, the responsibility of managing outages, repairs, and hardware issues falls onto the organization itself and not the vendor.

The second type of service, PaaS, gives the vendor the capability to control the runtime, middleware and the OS of the system. In this case, the organization has to handle only applications and data. PaaS is primarily developer-focused and can give users a shared platform for application development and management. That is an important DevOps component, with no need to build and maintain an infrastructure usually associated with this process.

Last but not least, a SaaS is a type of service that delivers a software application, which the vendor manages, to its users. Naturally, SaaS apps are web applications or mobile apps that users can access through a browser. A well-known example of this is Office 365 offered by Microsoft, where Microsoft controls the applications, data, runtime, middleware, OS, virtualization, servers, storage, and networking.

Let's delve into the different types of cloud. The first type is the public cloud, which is an environment created from IT infrastructure not owned by the end user. Some of the largest cloud providers are **Amazon Web Services (AWS)**, Google Cloud, and Microsoft Azure.

Traditionally, public clouds always run off-premises, but today's **Public Cloud Providers** (**PCPs**) have started offering cloud services on clients' on-premise data centers. This has made location and ownership distinctions obsolete.

Regarding public clouds, they can be used by abstracting the bare-metal IT infrastructure used by cloud providers and being sold as IaaS, or they can be developed into a cloud platform sold as PaaS.

The next type of cloud is the private cloud. Private clouds are defined as cloud environments, solely dedicated to a single end user or group, where the environment runs behind the end user or group's firewall. All clouds become private clouds when the underlying IT infrastructure is dedicated to a single customer with isolated access. This type of cloud is not just sourced from on-premises IT infrastructure, but organizations are also now building private clouds on rented, vendor-owned data centers located off-premises, which makes any location and ownership rules obsolete. This has led to many subtypes of private cloud, including the following:

- Managed private clouds
- Dedicated clouds

In the first subtype of a private cloud, customers create and use a private cloud that is deployed, configured, and managed by a third-party vendor. Those clouds are a cloud delivery option that helps enterprises with understaffed or underskilled IT teams to provide better private cloud services and infrastructure.

The second subtype of a private cloud, a dedicated cloud, is a cloud within another cloud. One way to have a dedicated cloud is to place it on a public cloud or a private cloud. For example, an accounting department can have its own dedicated cloud within the organization's private cloud.

The following type of cloud is a hybrid cloud. This is an ostensibly single IT environment created from multiple environments, connected through **Local Area Networks** (**LANs**), **Wide Area Networks** (**WANs**), **Virtual Private Networks** (**VPNs**), and/or APIs.

Their characteristics can be complex, and the requirements can differ, depending on which organization deploys such a cloud. For example, a hybrid cloud may need to include a bare-metal or virtual environment connected to at least one public cloud or private cloud. Every IT system can become a hybrid cloud when apps can move in and out of multiple separate yet connected environments. At least a few of those environments need to be sourced from consolidated IT resources that can scale on demand. Last but not least, all these environments need to be managed as a single environment using integrated management and orchestration platform.

There is also the multi-cloud option, which is made up of more than one cloud service, from more than one vendor. All hybrid clouds are multi-clouds, but not all multi-clouds are hybrid clouds. Multi-clouds become hybrid clouds when multiple clouds are connected by some form of integration or orchestration.

Having multiple clouds is becoming more common across organizations that seek to improve security, availability, and performance through an expanded portfolio of environments while reducing provider dependencies.

Having learned more about different cloud technologies can allow any organization to think before taking on a cloud technology.

Public clouds tend to have a wider variety of threats due to multitenancy and numerous access points. Public clouds tend to split security responsibilities. For instance, infrastructural security can be the provider's responsibility, as we saw from the different types of services, but workload security can be the tenant's responsibility.

Regarding private clouds, they tend to be more secure because the workloads usually run behind an organization's firewall, but that all depends on how strong an organization's security is.

A hybrid cloud is made up of the best features of every other environment where users and admins can minimize data exposure by moving workloads and data across environments, based always on compliance, audit, policy, or security requirements.

Having mentioned the different cloud technology types, now it's time to review what a blue team can do to protect a cloud environment. The most vital method that the blue team can use is monitoring and vigilance. Blue teams need to use telemetry on the cloud services to see whether they are used correctly by the organization or whether the cloud services have been breached by an external actor. This is because, usually, blue teams don't know what other users are using the public cloud – for example, they need to ensure that the instances their organization is using are not affected by other users of the public cloud. There has been a huge number of attacks on public clouds that have affected multiple businesses. A common example is overloading a cloud instance. Even though those are easy to handle by the vendors themselves when they happen, with users just moved to different, unaffected parts of the cloud, when the service offered is IaaS, workloads are not the provider's responsibility. This can become a huge nightmare for businesses, as the damage caused can be immense. Cloud-based providers generally charge the customers for cloud usage, so an unpredicted increase in workload can be costly. Additionally, the unavailability of services can cause damage to infrastructure, systems, and operations. The only way to prevent such an attack is to have the blue team monitor those instances and for them to have the authority to terminate any instance that might cause harm to the IT environment.

Let's take our castle example. The king has made an alliance with a foreign kingdom and started sharing land between the two kingdoms. The arrangement made is that the two kingdoms will share the spoils of the land by dividing their outcome evenly. The ones who will monitor this arrangement are the blue knights from both kingdoms. The blue knights of any of the two kingdoms have noticed that more resources are taken by the other allying kingdom. The blue knights make sure to stop the arrangement by informing the king of their decision.

Let's take the hospital example. The blue team in the hospital has made an arrangement to install new applications for the staff of the hospital. They installed SaaS applications, thus making sure that those applications are secure and ready for distribution among all the departments of the hospital. Next, they created a PaaS for the creation of the website for the hospital by the developers. Last but not least, for all the departments, the data was processed with IaaS technology. The blue team, of course, needs to monitor all those different cloud environments because they are connected, creating a multi-cloud hybrid environment. They placed web-application firewalls for the PaaS system, encrypted all the data at rest for all the systems, patched consistently, and, last but not least, managed access to all the different cloud environments.

In the next section, we will look at the other side of security, which is insider security. Are we sure we can trust our users?

Defense planning against insiders

When thinking about defense strategy, it is not always apparent that the workplace where we work is parallel to a battlefield, and therefore we can't predict certain types of behavior in the workplace. Organizations have been struggling with the notion of trust for many years, and what they have found out is that commitment to the company's goals and objectives should be rewarded. Certainly, rewarding employees with bonuses is one of the ways to prevent insider threats. However, if employees are not trained properly, then the issue is bigger than we realize. We see many mistakes made by employees that lead to insider threats.

An insider threat is a malicious threat to an organization that has, as its threat source, the people inside the organization, employees, former employees, contactors, or business associates, who have access to inside information concerning the organization's security practices, data, and information systems.

During the COVID-19 pandemic, there was a period when employees were fired or resigned. Scholars call this the Great Resignation. During that time, it was noticed that when employees were fired or resigned, they took away with them data that was important to the organizations they were working for, in order to get back at the organization they resigned or got fired from, but oftentimes, the data could be taken unintentionally, with a careless employee leaving a business with important information. For example, an employee leaving a company may decide to keep a USB stick with confidential or business-related information.

Hence, it is important to understand why organizations are not prioritizing insider threats. Most organizations do not have the budget to prepare for internal threats or lack internal expertise, but there are other reasons as well. In regard to those, firstly, many organizations do not perceive insiders as a substantial threat, and they may also claim that their organizational indifference to insider threats is due to internal factors, such as a lack of executive sponsorship. In fact, many of those organizations do not have an insider risk management strategy or policy, and a majority do not have a dedicated insider threat team.

This all means that organizations are woefully underestimating the seriousness of insider threats. Imperva, a company that analyzes insider threats, has noted that the biggest data breaches of the years 2016–2022 found 24% of those breaches were caused by human error or compromised credentials.

Despite the investment that organizations have placed in cybersecurity, they mostly focus on protecting themselves from external threats than paying attention to the threats lurking inside their own internal networks. Insider threats are harder to detect, especially during the COVID-19 pandemic when everyone was working from home and people were not meeting in the office anymore, so the blue team couldn't do its internal threat prevention or detection, making those threats invisible to regular security solutions, such as firewalls or internal detection systems. The lack of visibility into insider threats created a significant risk to the security of an organization's data.

Organizations, to prevent these threats, facilitated periodical manual monitoring, or auditing of employee activity and encryption of the data used by the employees daily. Many organizations also trained employees to ensure they were compliant with data protection policies. Even though those measures were taken, breaches and other InfoSec incidents are still occurring, and many organizations stated that their employees devised ways to circumvent their data protection policies.

It is really important that organizations add insider risk to their overall data protection strategy. An insider threat detection system must be diverse, combining several tools and methodologies to not only monitor insider behavior but also filter out false positives from the large amount of alerts coming through a system. Also, the protection of a company's intellectual property starts at the data layer; a wide-ranging data protection plan must include a security tool that protects the data layer, but that is not enough.

It is time to get to a solution. The steps that the blue team must take in order to have better protection against insider threats are as follows:

- Gain stakeholders' buy-in to invest in an insider threat prevention program. The CISO can facilitate this step by speaking to the board members at their meetings.

- Follow zero-trust principles to address insider risk. As mentioned previously, zero trust can be a great measure to prevent risks like these but usually only for bigger organizations. For smaller teams, building trust is really important.

- Build a dedicated function to address insider risk. The blue team has to divide its functions between outsider protection and insider protection. Of course, for smaller teams, the blue team can be used, but for bigger organizations, there has to be a dedicated group that takes care of insider threats.

- Create processes for an insider risk program and follow them. Strict policies should be implemented and followed. Every investigation should be treated as if it will end up in court, and policies must be applied consistently.

- Implement a comprehensive data security solution. A complete solution goes beyond **Data Layer Protection** (**DLP**) to include monitoring, advanced analytics, and automated response to incidents, thus preventing unauthorized, accidental, or malicious data access. The technologies deployed by the organization must support its objectives and the goals of the blue team, taking into regard the processes created in the last step and the mandate of the insider risk function.

Let's take an example. A team of developers will create a website for an organization, and they set backdoors into a system so that they can do their work better and faster. What the blue team noticed when they evaluated the system while the development work was underway was that those backdoors had a 10-digit PIN. That PIN was easily brute-forced by the red team, so the blue team set some defensive mechanisms in place for the developers, such as multi-factor authentication, better password strength, and username authentication, to see which developer is doing what on the system and monitor their progress.

If the blue team hadn't planned this and the project was completed and the backdoors left on the system, then there would have been chaos because a 10-digit PIN is easily penetrable by attackers, and they would have blamed the developers who left the backdoors open. That is why this is considered an insider threat.

In the hospital example, the HR person who changes the data in the HR operations database had a really strong migraine, and they were called by a fake number that was apparently, their boss ,who asked them to send data to their personal email. Certainly, that was a breach of company policy, but the employee was feeling really bad because of their migraine and complied with the attacker's demands without thinking. Afterward, the personnel data was leaked to the internet, and hackers managed to get the employees' private data, thus managing to blackmail the employees of the hospital. That was an example of employee neglect and blackmail.

In the castle example, the king incurred taxes on all citizens of his kingdom and many of the people wouldn't pay the taxes, so they revolted and attacked the king, while he was resting in his castle. This example is a disgruntled employee example. It can happen in almost any organization, which is why employers and blue teams should be aware when an employee is not feeling well or is not happy about something. The employer needs to offer help in that situation.

In the next part, we will be referring to the different responsibilities or roles that the blue team has to take in order to implement a defense strategy.

Responsibilities in blue team operations

Blue teams operate in many information security sectors. Let's look at them one by one.

The CISO, who is the lead of the blue team, always stands in every sector as the product owner to those sectors:

1. **Firstly, the infrastructure sector**: Here, we see network security personnel whose prime function is to prevent attackers from disrupting operations and getting data out of payloads. This personnel includes firewall monitoring analysts, network operation analysts, network engineers, and cybersecurity engineers. Generally, their goal is to protect the network and stop interruptions to the operations of that network or infrastructure.

2. **Secondly, the system sector**: Here, we have operational personnel who handle the systems of an organization, making sure that everything works as planned and no one strays away from normal operations. Roles deriving from this sector include operations analysts, systems analysts, and cybersecurity analysts.

3. **Thirdly, the endpoint sector**: Here, we have monitoring personnel and other types of information security analysts who work to protect end users from outside threats. Roles deriving from this sector can include service desk personnel who report to cybersecurity analysts, who will track their reports.

4. **Fourthly, the cloud sector**: In this sector, there are people who handle the operations of the cloud, making sure that everything works as it should. This can include roles such as cloud security analysts, infrastructure analysts, and cybersecurity analysts.

5. **Lastly, the internal security sector**: In this sector, the entire function works toward one goal – the prevention and mitigation of insider threats. This can include roles such as human relations analysts, security awareness officers, HR analysts, and cybersecurity analysts.

Generally, cybersecurity analysts are the members of the blue team that should have an understanding of multiple sectors so that they can move around each year to analyze each sector separately.

Summary

In this chapter, we looked into the regular schedule of a blue team and their usual responsibilities, including system security, endpoint security, cloud security, and insider defense planning. We also covered the blue team's strategy and why that is important. We looked at some analogies, a medieval kingdom and a hospital, to understand this better, and always remember to make such analogies whenever we need to explain what the blue team is doing at any specific time.

In the next chapter, we will be covering threats and how those threats are being monitored by the blue team, what will they notice, and what will they not notice if different circumstances occur.

5
Threats

In this chapter, we will learn more about cyber threats, how they are accomplished, and what blue team members will come across in their work. You will learn about how different threat actors act, and what you should expect from them. It is worth noting that there are innumerable infrastructure infiltrations, data breaches, spear phishing attacks, and brute-force attacks nowadays. There is a wide variety of cyber threats, and they don't discriminate between individuals and organizations.

In this chapter, we will be dealing with the following:

- What are cyber threats?
- The Cyber Kill Chain
- Internal attacks
- Cyber attack actors
- The impacts of cybercrime

What are cyber threats?

A great blue team member should be able to get into the mindset of an attacker, whether they are external or internal. They must be able to predict how those attackers would react if they were placed inside a defensive block. A blue team should be proactive, defending and attacking at the same time. In this section, we will try to get into the mind of an attacker – how they think, what their goals would be, and why they want to accomplish those goals.

The word *cyber* was first used in the 1950s to describe a scientific field called cybernetics, which studies how machines and animals are controlled and moved. Beyond this, the term *cyber* refers to a computerized system.

A new term was coined in the 1990s – cyberspace – a physical space developed behind the electronic activities of computing devices that some people believed existed.

Data theft, data damage, and damage to digital well-being are all cyber threats that threaten the stability and well-being of an enterprise. Among the many types of cyber threats are data breaches, computer viruses, **Denial-of-Service (DoS)** attacks, and numerous other types.

A cyber threat is any activity that can cause serious harm to a computer system, a network, or any other digital assets of an organization or an individual that has the potential to cause serious harm to them. A report by Techopedia refers to cyber threats as attacks designed to exploit real vulnerabilities in systems and networks to gain access to them. Among the many threats that exist in the cyber world, there are trojans, viruses, hackers, and backdoors, to name a few. As a single cyber threat can involve multiple exploits, it is usually more appropriate to refer to it as a *blended cyber threat* since one threat can contain multiple elements. Using phishing attacks, a hacker can get information from a company's computer system and then use that information to break into the network.

Similarly, cyber threats can be defined as any circumstances or events with the potential to infiltrate a computer network, disrupt an **Information Technology (IT)** asset, damage intellectual property, steal or damage any form of sensitive data, or gain unauthorized access to a network. These threats can be posed by trusted employees within an organization or by unidentified external parties in remote locations.

Our lives are deeply affected by cybersecurity threats, and it is not an exaggeration to say that. Cyber threats can lead to blackouts, the failure of military equipment, or the breach of national security interests. These attacks can disrupt computer and phone networks, paralyzing systems and preventing data transmission. As well as stealing sensitive, valuable data, they can also steal personal information across the globe, such as medical records.

In the end, the intent and potential impact of the attacker matter. Many cyber attacks are mere nuisances, but some get quite serious, even posing a threat to human life.

Every new technology brings new threats alongside its intended benefits. This is especially pertinent when implementing IT systems within an organization as part of a business process. It can help improve an organization's workflow, but at the same time, it can be a target for cyber attacks. A simple goal of cybersecurity within an organization is to prevent unauthorized and unexpected access to their computer systems and data and any illegal changes being made to them, as well as protection from other potential threats.

During the Covid pandemic, cybercrime has been predicted to cost companies worldwide an estimated $10.5 trillion annually overall by 2025, up from $3 trillion in 2015. In terms of growth, we've seen a rate of 15% increase year over year. Cybersecurity Ventures reports that cybercrime represents the greatest transfer of economic wealth in history. Many companies have made significant digital transformations and now use online services for advertising, sales, customer outreach, and employee recruiting. Considering the importance of these technological operations, corporations must be safeguarded from outsiders who might sabotage the infrastructure of a company or its data by any means necessary.

Is there anything in particular that makes an organization vulnerable to cyber attacks? Almost all aspects of networking systems can be compromised, whether the hardware, any IT-based services, customer data, as well as other sensitive information when it comes to network systems. In addition to the cyber attacks in the hardware, network, or IT, they can also include thefts and illegal takeover of computers, mobile devices, and IT systems and websites on a remote basis, and the theft of company data stored on third-party systems, such as cloud services or partner companies.

Most organizations are aware that cyber attacks are a threat, but they are not always aware of the losses that may result from a cyber attack or its effects. Cyber attacks can seriously affect your organization because they can attack many areas simultaneously. There are several risks associated with this kind of event, among those being the serious loss of stored funds, the cost of recovery and replacement, the diminished reputation of the company, the need to pay fees, and the fact that the system may have been linked to another company.

Since technology is becoming increasingly sophisticated, in many cases, the skills of criminals are more sophisticated than those of the security professionals within an organization. This is observable given that most attacks are successful. There is a great need for qualified and well-trained engineers, as cybersecurity plays an increasingly important role in society today.

Organizations must hire people with functional skills to secure their networks and protect systems, computers, and data from being attacked, damaged, or accessed by unauthorized individuals.

The Cyber Kill Chain

The Cyber Kill Chain is an adaptation of a military kill chain. It is a step-by-step process used to identify and contain enemy activity – a systematic approach to identifying and eradicating it. The Cyber Kill Chain can be seen in the following diagram:

Figure 5.1 – The Cyber Kill Chain

The Cyber Kill Chain is a framework that describes the various stages of several common cyber attacks and, along the same lines, the various points at which an information security team can prevent, detect, and intercept these attacks. Lockheed Martin originally developed the Cyber Kill Chain in 2011.

Seven sequential steps were involved in Lockheed Martin's original Cyber Kill Chain model.

Phase 1 – reconnaissance

Reconnaissance is when a malicious actor identifies a target and investigates vulnerabilities and weaknesses within the network that may allow them to exploit the target. To accomplish this goal, the attacker may gather passwords or other information, including email addresses, user IDs, physical locations, software applications, and **Operating System** (**OS**) details, all of which may be gathered to use in phishing or spoofing attacks, especially when the attacker has physical access to a system. Generally, the more valuable information the attacker can collect during the reconnaissance phase, the more sophisticated and convincing the attack will appear to be, and, therefore, the higher the chances of the attack succeeding will be.

Social engineering attacks are common reconnaissance techniques that attackers use to exploit vulnerabilities in people and systems and can include the following:

- **Pretexting**: Under pretenses, the attacker attempts to pressure the target into providing information under similar circumstances to baiting. It is common for the perpetrator to pose as someone with authority, such as an IRS official or a police officer, to force the victim into complying with their demands.

- **Phishing**: To trick the recipient, the attacker sends an email that appears to come from a trustworthy source. Email phishing often involves sending fraudulent emails to many recipients, but targeting specific groups of recipients is possible. An example of spear phishing would be sending a targeted email to a specific recipient, while a whaling attack would target high-value individuals or CEOs. For example, the attacker sends an email in which they have placed a link that indicates it is from a bank with which the user has an account. The website is a copy of the login site of the bank. The user may not notice the subtle differences in the website, so they enter their bank login details. Now, the hacker has gained access to the bank account of their target. All the money in it has been moved to a Swiss bank account as a donation. This is an example of an attack. The motive, in this case, was to get the account details and access the money in the account. A blue team member can interfere in this process in various ways. Firstly, if the reporting process of the organization is good, they will know of the email's existence and thus stop the attacker from gaining the prime advantage. Say that the user clicks on the link and provides their e-bank account details. Banks have implemented measures to stop attackers from accessing all the money in an account by putting multi-factor authentication in place, for example, to stop attackers in their tracks – or so they think. An attacker using this method has been targeting their victim, meaning they may have found the user's mobile phone information and copied it so that they can receive the messages that the user's phone receives.

- **Voice phishing** (**vishing**): The imposter can access the target's system or get them to disclose sensitive data on the phone. Anyone can be targeted by vishing, but it is typically used against older individuals. This has been noted in many different attacks, including automated phone calls and calls from so-called insurance providers, which can cause huge issues. This threat originates from advertisements online in which the end user has revealed their personal data (mobile phone, first name, last name, and so on).

- **SMS phishing** (**smishing**): The attacker uses text messages to deceive the victim. For example, a common attack will include an SMS that tells the user something like the following: *Post Office: Your driver Sasha tried to deliver your parcel today. To reschedule and track your parcels, visit http://local-serviceassist.com.* If the user follows this link, they are led to a website with **malicious software** (**malware**) on it and their phone is infected.

- **Supply chain attacks**: Attacks on supply chains pose a new threat to software developers and vendors. They intend to infect legitimate applications and distribute malware through various means, such as source code, build processes, or software update mechanisms. There is an increase in attacks that target non-secure networks, server architectures, and coding techniques. The attackers target them to compromise build and update processes, modify source code, and hide malicious content in the code. The problem with supply chain attacks is that compromised applications are typically signed and certified by trusted vendors, but attackers misuse them for bad purposes. An example of a supply chain attack is when the software vendor, as opposed to the end user, is unaware that their application or update has been infected with malware because they don't know when it was infected. The result is that malicious code is run in the same manner and with the same degree of privileges as the compromised application on which it is based. Several types of supply chain attacks exist, including the following:

  - Building tools and development pipelines that have been compromised

  - Code signing procedures that have been compromised, or developer accounts that have been compromised

  - Malicious code is sent to systems as an automated update to hardware or firmware components through the internet

  - Devices that have malicious code pre-installed into them

- **Man-in-the-Middle (MITM) attacks**: MITM attacks occur when communication between two endpoints, a user and an application, is intercepted. A hacker may be able to eavesdrop on communications between parties, steal sensitive data, and impersonate each party involved. Examples of MitM attacks include the following:

  - **Wi-Fi eavesdropping**: Attackers set up Wi-Fi networks pretending to be legitimate actors, such as businesses, so that users can connect to them as if they were legitimate actors. Using the fraudulent Wi-Fi network, the attacker can monitor the users' online activity and intercept sensitive information, such as payment card information and login credentials.

  - **Email hijacking**: Using a spoof email address, a perpetrator can get users to provide the attacker with sensitive information, such as their credit card number or bank account information, misleading them into providing access to money. A user is led astray by instructions they believe come from a bank, but in reality, they are being sent by an attacker.

  - **Domain Name System (DNS) spoofing**: The DNS of a legitimate website is spoofed, causing the user to be redirected to a malicious website posing as a legitimate website. Cyber attackers may steal users' credentials or traffic diverted away from legitimate websites.

  - **IP spoofing**: A website is connected to an IP address, and users can be deceived into thinking they are interacting with a real website if an attacker spoofs the IP address.

  - **HTTPS spoofing**: HTTPS is generally regarded as a more secure version of HTTP, but a malicious site can also masquerade as a legitimate website using this protocol. To conceal the website's malicious nature, the attacker uses HTTPS in the URL.

Phase 2 – weaponization

At the point of weaponization, the attacker develops an attack vector that can exploit a known vulnerability, such as remote access malware, ransomware, viruses, or worms, to gain access to the target computer. As this process continues, the attacker may also set up backdoors so that they can continue to access the system after the point of entry that they were using is identified and closed by the network administrator.

A malware attack can be classified as a virus, worm, trojan horse, spyware, or ransomware. It is the most common form of cyber attack. An attacker infiltrates a computer system via a link on an untrusted website or email or unwanted software from a website or email being downloaded by accident. In their analysis before the reconnaissance phase, the program has been established as being deployable on the target system, as well as able to collect sensitive data, manipulate and block access to network components, destroy data, and shut down the entire system.

Malware attacks fall into the following categories:

- **Viruses** are pieces of code in an application or file that have malicious properties, and the malicious code executes when the application or file is run and delivers its payload without the need for command and control. One example of this type of malware is a user receiving an email with the subject *Love letter for you*. The user opens the email and reads *My love letter to you*. The email has an attachment, `LoveLetter.txt`. The user is not smart enough to guess that a hacker has hidden a virus in that `.txt` file with a hidden `.vbs` extension. When the user clicks on the file, Notepad does not open – a program starts running behind the scenes, deleting all photos and songs and copying itself in every photo's and song's place. Then, it copies itself onto the user's contact list and spreads itself using their email. This is a direct-action virus. This virus does not need a hacker's attendance in order to execute a payload. This is another way in which we can see that people are not needed on hand for crime to be done – software can act automatically. The question is why the hacker wanted to delete photos. Only the hacker can answer that question. However, this well-known I LOVE YOU/Love Bug/Love letter for you virus originated in the Pandacan neighborhood of Manila in the Philippines on 4 May 2000. The outbreak was later estimated to have caused $8.7 billion in damages worldwide. Within 10 days, over 50,000,000 infections had been reported. The damage cited was mostly the time and effort spent getting rid of the infection and recovering files from backups, if any existed. To protect themselves, most large corporations decided to completely shut down their mail systems. At the time, it was one of the world's most destructive computer-related disasters ever, without being attacker-driven.

- **Worms** are standalone malware computer programs that when executed, replicate themselves in order to spread to other computers. Usually, attackers deploy worms in order to spread a trojan horse, virus, or another type of malware. It is possible to gain access to an OS through malware exploiting software vulnerabilities and backdoors. In addition, the worm can carry out attacks such as **Distributed Denial-of-Service (DDoS)**.

- **Trojan horses** are innocuous-looking programs or apps that can hide malicious code or software within themselves, whether apps, games, or attachments in an email that pose as innocent. By downloading a trojan, an unsuspecting user allows it to control their device, and until it is detected, it delivers its payload. Usually, they are automatic, but there are cases where the attacker uses the trojan horse to carry out a DDoS attack against another system. Those are called botnets, and an attacker can use those botnets to gain computing power to deliver other types of attacks.

- **Ransomware** is malware that encrypts the data in a system. In the meantime, users or organizations cannot access their systems or data because data or systems have been encrypted. The attacker typically makes a ransom demand in exchange for a decryption key that enables full access to the system again. However, paying the ransom does not ensure full access or functionality returns and it doesn't ensure that the data won't be published online. The way to deal with ransomware is discussed in *Chapter 10, Incident Response and Recovery*, of this book. For example, looking at a time closer to 2022, during May 2017, a ransomware attack became prominent in the **National Health Service (NHS)** of the UK, which was not able to handle such an attack. It goes to show that even great organizations such as the NHS are vulnerable to ransomware. Computers running the Microsoft Windows OS were targeted by encrypting their data and demanding ransom payments in the Bitcoin cryptocurrency in return. Researchers classified this ransomware as a computer worm when talking about how it spread. Vulnerable systems were scanned using its transport code, the EternalBlue exploit was used to gain access, and the DoublePulsar tool was installed and executed a copy of itself. WannaCry versions 0, 1, and 2 were created using Microsoft Visual C++ 6.0. EternalBlue is an exploit developed by the **National Security Agency (NSA)** that took advantage of a vulnerability in Microsoft's **Server Message Block (SMB)**. This vulnerability existed because the **SMB version 1 (SMBv1)** server in various versions of Microsoft Windows mishandled specially crafted packets from remote attackers, allowing them to remotely execute code on the target computer. This allowed the malware to execute code that allowed it to enter the infected machines. DoublePulsar is a backdoor implant that was developed, again, by the NSA. This tool works in kernel mode, granting cybercriminals a high level of control over a computer system. Once installed, it uses three commands, `ping`, `kill`, and `exec`. The latter can be used to install malware on the system as it did for the WannaCry malware, which used this backdoor mechanism to move to another vulnerable target. The damages were huge. The attack began on Friday 12th May 2017, and within just a month of operation, as of 14th June 2017, after the attack had subsided, a total of 327 payments totaling $130,634.77 (51.62396539 XBT) had been transferred. The experts quickly discerned that paying the ransom did not decrypt the files, so they suggested against paying the ransom. This has become a saying in the security world, "*never pay the ransom they demand of you; it won't decrypt anything.*" However, getting around the ransomware problem is not an easy feat for any blue team. In this case, experts have deduced that it wasn't a matter of a zero-day attack, but a matter of patching. The affected computers were not patched or were legacy machines, so changing those machines or patching them was the solution to all their problems.

- **Cryptojacking** happens when, unknown to the victim, an attacker installs software onto their device and begins generating cryptocurrency with their computing power. Cryptojacking kits can slow down affected systems and compromise system stability. Cryptojacking kits are usually used along with trojan horses that have unknowingly been installed by an unsuspecting user.

- **Spyware** is a type of malware that can get to an unsuspecting victim's data, including payment details and passwords, which are then stolen by a malicious actor. A malware program such as this can easily affect a computer's desktop browser, a mobile phone, and a desktop application, and all the shared data that the user has placed in each of those devices will be compromised.

- **Adware** is a type of malware that tracks a user's browsing activity to determine their interests and behavior patterns, allowing advertisers to send targeted advertisements. While adware may be related to spyware, it does not require users to install software on their computers, nor is it always used maliciously. However, it can compromise their privacy if used without their consent.

- **Fileless malware** is a type of malware with which users don't install software but some native files are modified to enable malicious functions, such as WMI and PowerShell. Since the compromised files are recognized as legitimate, this stealthy attack is hard to detect (antivirus software cannot detect it).

- **Rootkits** are malware that provides remote access to computer systems, applications, firmware, kernels, or hypervisor software, and are injected into computer systems to control them remotely. A computer becomes completely controllable as soon as the OS starts, allowing the attacker to install additional malware and take complete control of the computer.

Phase 3 – delivery

The intruder launches the attack in the delivery step of the attack process. The specific steps that need to be taken will vary depending on what type of attack they intend to conduct. A hacker may, for example, send an email attachment or link that will spur the user to click on it to advance their plan. Social engineering techniques may be combined with this activity to increase the campaign's effectiveness.

Phase 4 – exploitation

There is then an exploitation phase in which the malicious code will start to execute on the victim's machine, exploiting the vulnerabilities that are there in the systems. An example of this phase could be the Log4Shell vulnerability. A vulnerability was found in Log4j, an open source logging library commonly used by apps and services across the internet. If left unfixed, attackers can break and enter systems, steal passwords and authentication logins, extract all data on a system, and infect networks with malware. Developers use Log4j worldwide across software applications and online services, and the vulnerability requires minimum know-how to exploit. This makes Log4Shell potentially one of the most severe computer vulnerabilities in years.

Log4j is a collection of useful building blocks that developers can use. Modern software can be large, powerful, and complex. Rather than a single author writing all the code themself, as was common decades ago, modern software creation involves large teams, and that software is increasingly made up of *building blocks* pulled together by the team rather than entirely written from scratch. A team is unlikely to spend weeks writing new code when they can use existing code immediately. Log4j is one of the many building blocks that are used in the creation of modern software. It is used by many organizations to do a common but vital job. This is called a *software library*. Log4j is used by developers to keep track of what happens in their software applications or online services. It's basically a huge journal (or log) of the activity of a system or application.

Log4Shell works by abusing a feature in Log4j that allows users to specify custom code for formatting a log message.

Unfortunately, this kind of code can be used for more than just formatting log messages. Log4j allows third-party servers to submit software code, which can perform many actions on a targeted computer. This can open the door for nefarious activities such as stealing sensitive information, seizing control of a targeted system, and allowing the attacker to slip malicious content such as malware to other users that communicate with the server affected.

Phase 5 – installation

After the exploitation phase is completed, you can expect to see your victim's system infected with malware or another attack vector shortly afterward. Having entered the system, this is the crucial beginning of the attack life cycle because the threat actor can now take control of the system, having entered it.

Phase 6 – command and control

A command-and-control attack occurs whenever an attacker manages to make an identity or device within a target network be infected with malware so that it can be used for the remote control to infect the device or devices connected to that device. In addition, the attacker can also use this stage to move laterally through the network, gain access, and create additional points of entry for the future (backdoors). As mentioned previously, botnets are a type of command-and-control technique with which attackers use the computers they have placed in the botnet to attack larger systems; for example, they can attack, using the botnet, a well-known website that they hope to perform DDoS against, make changes to the frontend, deface it by using cross-site scripting attacks, or perform other injection attacks on that website.

An **injection attack** inserts malicious code directly into the code of a web application by exploiting its vulnerabilities. If an attack is successful, sensitive information may be exposed, a DDoS attack can be launched, or the whole system may even be compromised. Injection attacks have different types:

- **SQL injection** is a type of attack in which attackers enter **Structured Query Language (SQL)** queries into end user input channels, such as web forms or comments. When an application is vulnerable, the attacker will be able to inject SQL commands into the query and extract a user's data from the database. SQL injection is a vulnerability of web applications that use SQL databases. There is also a new variant of this attack that targets database structures without a relational structure – **NoSQL attacks**.

- **Code injection** is when, depending on the application's vulnerability, an attacker injects code into it to cause damage. In the course of executing the application, the malicious code is executed by the web server.

- **OS command injection** is an attack in which an attacker exploits a command injection vulnerability and can input commands that the OS will execute in response. The attacker can exfiltrate OS data or take over the whole system this way.

- **LDAP injection** is an attack in which an attacker enters a character into an LDAP query to alter it. Unsensitized LDAP queries make a system vulnerable. Since LDAP servers can store the credentials of all of the users in an organization, these attacks can be extremely serious.

- **XML External Entities (XXE) injection** is an attack on a network that uses specially crafted XML documents. Instead of exploiting unvalidated user inputs, this attack vector exploits an inherent vulnerability in legacy XML parsers, which makes it different from other attack vectors. Using XML documents, attackers can traverse paths, execute code remotely, and perform **Server-Side Request Forgery (SSRF)** attacks.

- Lastly, **Cross-Site Scripting (XSS)** is an attack in which during the attack, the attacker enters a string of text that contains malicious JavaScript code in the text field. When executing this code, it redirects the target to a malicious website or steals cookies to hijack the user's session. Applications that fail to sanitize user inputs to remove JavaScript code are vulnerable to XSS.

Phase 7 – actions on objective

A hacker takes several steps during this stage to accomplish their intended objectives, including data theft, destruction, encryption, and exfiltration. At this stage, it would be beneficial to mention DDoS attacks and how they are accomplished by attackers to disable the availability of a system, even when we don't know whether that was the attacker's prime objective or not.

A DoS attack overwhelms a target system with excessive traffic, preventing it from functioning normally. The term DDoS refers to an attack that involves multiple devices. Techniques used in DDoS attacks include the following:

- **HTTP flood DDoS** is a technique via which an attacker overwhelms a web server or application with legitimate HTTP requests. Generally, this technique forces an entire target system to allocate all available resources to each request without requiring high bandwidth or malformed packets.

- **SYN flood DDoS** is a technique in which the SYN-ACK sequence between servers and users is exploited. To exploit this sequence, attackers can send SYN requests but not respond to SYN-ACKs from the host, tying up server resources.

- **User Datagram Protocol (UDP) flood DDoS** is a technique in which UDP floods random ports on a remote host with packets. This technique results in the host searching for applications running on those ports and responding with *Destination Unreachable* packets, which consume a lot of host resources.

- **ICMP flood** is a technique that occurs when an influx of ICMP echo request packets overwhelms the target, consuming both the target's inbound and outbound bandwidth. The systems slow down because the servers cannot handle each request, so each request is responded to by an ICMP echo reply packet, but the system cannot handle the rate of requests.

- **NTP amplification** is a technique where an attacker, using NTP, can send large amounts of UDP traffic to a targeted server. NTP is a protocol designed to keep track of time. An attacker can execute high-volume, high-bandwidth DDoS attacks by exploiting NTP servers with a query-to-response ratio of 1:20 to 1:200.

Although DoS and DDoS attacks are and will always be out there, organizations should be aware of the types that exist in order to prevent them because they can lead to bigger threats than just unavailable servers. This includes systems becoming compromised and blue teams seeing data exfiltration techniques.

There are many techniques for data leakage, but it is often done either physically or digitally. In most cases, it is carried out via email, such as phishing. Data can be employee info, customer databases, intellectual property, payment card information, **Personally Identifiable Information (PII)**, or other financial data.

Regardless of the form it takes, data leakage has serious consequences for businesses in every industry. Failure to mitigate data leakage and take as many preventative measures as possible can lead to the following:

- The loss of intellectual property and other sensitive information

- Expensive incident response processes

- Information misuse or abuse

- The violation of industry standards and regulations

- Lawsuits and other legal issues

- Reputational damage

Adequate protection is becoming an increasingly difficult task for security leaders for two main reasons. First, cybercrime has evolved from an individual act to an organizational endeavor. Attackers now have the budget, resources, and sophistication required to advance their data leakage methods. In addition, an organization's data infrastructure contains various tools – originally designed for legitimate data sharing – that can be used to get access to data. The most common data exfiltration techniques used by hackers are the following:

- **Phishing**, as we've explained before, is a prime technique that hackers use to exfiltrate data and to see whether users will bite.

- **Tailgating and piggybacking** are two techniques primarily used in the physical world. Tailgating epitomizes a situation in which an individual without the proper authorization closely follows an authorized person in a reserved area. The malefactor takes advantage of the moment when the authorized person opens the door with their badge and sneaks inside before the door closes. Piggybacking corresponds to a situation in which someone gains entry to a reserved area without obtaining permission by deceiving an authorized person.

- **A watering hole** is a computer attack strategy in which an attacker guesses or observes which websites an organization often uses and infects one or more of them with malware. Eventually, a user from the targeted organization will become infected. Attacks such as this look for specific information and thus hackers may only attack users coming from a specific IP address. This also makes the attacks harder to detect and research. The name is derived from predatory animals in the natural world who wait for an opportunity to attack their prey near watering holes.

- **Scareware** is a form of malware that uses social engineering to cause shock, anxiety, or the perception of a threat in order to manipulate users into buying unwanted software. Scareware is part of a class of malware that includes rogue security software, ransomware, and other scam software that tricks users into believing their computer is infected with malware, then suggests that they download and pay for fake anti-malware to remove it. Usually, the virus is fictional, and the software usually doesn't work or is malware itself.

- **Dumpster diving** is the search for treasure in someone else's trash. In the world of IT, dumpster diving is a technique used to retrieve information that can be used to carry out an attack or gain access to a computer network from discarded objects. Dumpster diving is not limited to searching the trash for obvious treasures, such as passwords or passwords written on sticky notes. Seemingly innocent information, such as a phone book, a diary, or an org chart, can be applied to help an attacker use social engineering techniques to gain access to a network. To prevent dumpster divers from learning anything of value from trash, experts recommend that businesses establish a disposal policy where all paper – including printouts – is destroyed in a cross-cut shredder before being recycled, all storage media is wiped, and all staff is trained on the risk of undetected trash. Discarded computer hardware can be a gold mine for attackers. Information can be recovered from storage media, including drives that have been improperly configured or deleted. This includes stored passwords and trusted certificates. Even without storage media, equipment may include **Trusted Platform Module** (**TPM**) data or other hardware identifiers that are trusted by an organization. An attacker can also use the hardware to identify the equipment manufacturer to create potential exploits. Medical and personnel records may have legal consequences if not disposed of properly. Documents that contain PII must be destroyed or the organization may be exposed to breaches and possible fines:

Dumpster diving

This entails combing through someone else's trash to find treasures - or in the tech world, discarded sensitive information that could be used in an illegal manner. Information that should be securely discarded includes, but is not limited to:

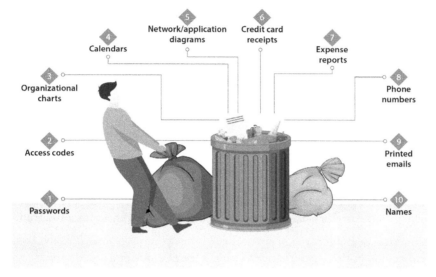

Figure 5.2 – Dumpster diving

- **Shoulder surfing** is a type of social engineering technique used to obtain information such as **personal identification numbers** (**PINs**), passwords, and other confidential data by looking over a victim's shoulder. Unauthorized individuals monitor keystrokes entered into a device or listen to sensitive information said aloud, which is also known as snooping or eavesdropping. This type of attack can also be undertaken in the digital world, meaning by hackers who record the whole login process without being physically present to shoulder-surf. In this case, attackers will use binoculars, tiny stealthy video cameras, or other devices that rely on optical technology to spy on their targets. The goal is to obtain information such as usernames and IDs, passwords, PII, and credit card numbers. **Pretexting**, as we explained before, is the attack before the main attack and can lead to spear phishing or other attacks of greater magnitude.

- **Baiting** is a type of attack in which a social engineer uses a false promise or reward to trap victims and steal their sensitive information by infecting their system with malware. The lures are very attractive and enticing, not to mention manipulative, and their ultimate goal is to infect your system and gain access to personal information. Attackers who practice baiting use physical devices and tempting offers that appeal to people's inquisitiveness or trap them to get what they want from their victims. In many ways, baiting is similar to phishing, but it differs from most social engineering attacks overall. This is because these attacks offer something for free that is pertinent to the victim.

- **Diversionary theft** is a cyber attack that is launched offline. In this attack, an attacker convinces a courier to pick up or drop off a package at the wrong location, deliver the wrong package, or deliver a package to the wrong recipient. Diversionary theft has since been adapted as a web-based scheme. The malicious actor steals confidential information by tricking the user into sending it to the wrong recipient. This type of attack often involves spoofing, which is a technique used by cybercriminals to disguise themselves as known or trusted sources. Spoofing can take many forms, including spoofed emails, IP spoofing, DNS spoofing, GPS spoofing, website spoofing, and spoofed phone calls.

- **Whaling** is a type of phishing attack that also uses personal communication to gain access to a user's device or personal information. The difference between phishing and whaling relates to the level of personalization. Phishing attacks are not personalized and can be replicated for millions of users while whaling attacks target one individual, usually a high-level executive. This type of attack requires significant research on the individual, which is usually done by examining their social media activity and other public behavior. This in-depth investigation results in a more sophisticated approach and a greater likelihood of success. Although whaling attacks initially require more planning and effort, they often have huge benefits, as the targets have access to high-value data or the financial resources needed to proceed with a ransomware attack.

- **Impersonation** is another type of cyber attack carried out with the malicious intent to steal confidential information. Cyber attackers do not use malware or a bot to commit cybercrime via impersonation attacks – rather, they use another powerful social engineering tactic. The attacker researches and collects information about the legitimate user through a platform such as social media and then uses that information to impersonate or pretend to be the original legitimate user. Impersonation attacks pose a security threat because they involve direct action that forces action without distinguishing between the authenticated user and the impersonated user. The nature of an impersonation attack makes it a very dangerous form of cyber attack, as users have the right to privacy. Impersonation attacks can be carried out using similarities to the identity of the original user, such as email identifiers. An impersonation attack can have many different types:

  - **Email impersonation**, as we've mentioned earlier, is when a hacker pretends to be a colleague, manager, or high-level supervisor using a fake or stolen email account. In contrast to mass email phishing attacks that end up in the spam folder, impersonation attacks (or spear phishing attacks) are highly sophisticated and targeted attacks. Email impersonation attacks often contain malicious links or images that can take the user to a compromised or malicious website that contains malware. Other attacks will use social engineering methods to trick the employee into revealing important data or transferring funds directly to the attacker.

  - **Cousin domain** impersonation attacks are when an attacker creates a false company site or email nearly identical to the official organization websites using the wrong domain codes. Usually, domain codes such as `.org`, `.net`, `.com`, or `.tech` will be used on the cousin websites, but the wrong domain code to falsify their emails will be used by a cousin domain attack. Completely copying the website design of a legitimate website can be a technique of attackers.

- **Email spoofing** uses emails with fake headers or sender addresses that appear to be legitimate emails. A scammer uses a fake email that's hidden behind a false header or envelope using a recognized name or title, thus successfully bypassing spam filters. These forged headers can trick people into trusting that a legitimate source sent a message if they only read the header and not the email address. In this attack, the *sender* field in an email header or envelope is modified by changing the **From** or **Return-Path** title fields, making it appear that a legitimate business or a friend is sending the email.

- An **Account Takeover** (**ATO**) is an attack in which cybercriminals log in to an account with stolen credentials such as usernames and passwords, often bought on the dark web and exfiltrated through data breaches, data leaks, or other brute-force attacks. If an account doesn't have multi-factor authentication, a hacker can successfully log into the specific email account and even use it for other ATOs and identity theft. Many people may use the same password for multiple sites, making it easier for cybercriminals to access multiple sites and accounts of that user. Using a compromised or stolen account, attackers can send phishing emails to other contacts in their email list, making it near impossible for victims to recognize this type of attack.

- **MITM attacks**, as we explained before, can be a huge hassle for blue teams if they are not identified quickly.

- **Smishing** and **vishing**, as explained before, are prime threats in the cybersecurity world, and blue teams should be able to make users aware of these threats and make this type of impersonation attack obsolete.

Over the last few decades, many information security professionals have identified an eighth step in the kill chain: **monetization**.

Cybercriminals use this phase to generate income from the attack by demanding ransom or selling sensitive information on dark web marketplaces, such as personal information or trade secrets.

Generally, the earlier a company can detect and stop a cyber attack during its life cycle, the less risk it will be exposed to. To remedy an attack that has reached the command-and-control phase, much more advanced remediation efforts are usually necessary, including an in-depth sweeping of the network and endpoints to accurately evaluate the extent and depth of the attack. As a result, organizations should identify and neutralize threats as early as possible in the threat life cycle to minimize the risk of an attack and the costs associated with resolving an incident, both of which are associated with an attack.

One of the things we should learn from this chapter is that attackers are out there and they are looking at our systems to exploit their vulnerabilities, so we should not let them do so. In the next section, we will discover how internal attackers might think and the motives they might have to betray an organization.

Internal attacks

Internal attackers are a real nuisance to blue teams. Between 2018 and 2020, there was a 47% increase in the occurrence of incidents involving insider threats.

This includes malicious data breaches and unintentional data loss. The latest research, from the Verizon 2021 Data Breach Investigations Report, advocates that insiders are accountable for around 22% of security incidents.

Why does this matter? Because these incidents cost organizations millions, are leading to breaches that expose sensitive data, and are especially hard to prevent.

There are also several different types of insider threats and the *who and why* behind these incidents can vary. According to one study, we see the following:

- Neglectful insiders are the most common vectors and account for 62% of all incidents
- Neglectful insiders who have their credentials stolen account for 25% of all incidents
- Hostile insiders are responsible for 14% of all incidents

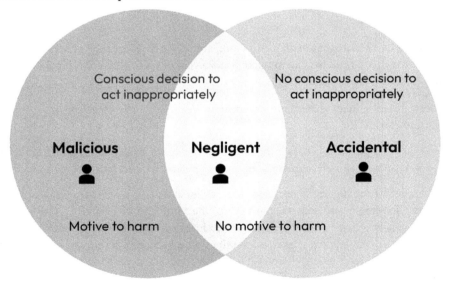

Figure 5.3 – Insider attacker types

Taking on all insiders at once is quite a feat but an organization's blue team should keep their eyes on the prize, as they say. Always talk with people. If they betray their company, the managers who didn't notice that their employee was acting strangely will be at fault. This applies to malicious and negligent employees who fall victim to their own bad attitude or psychology, or just a bad moment, and can cause serious damage to the organization's reputation and financial standing. If employees are not rewarded properly, then they become disgruntled and that is not good for any organization. We have already spoken about the solution to human error, which is training employees not to cause any accidental internal attacks. Beyond that, for malicious and negligent employees, the best solution is recognition, recognizing their work, appreciating them, and giving them the bonuses they deserve.

If there is no ulterior motive behind an internal attack, then recognition is really meaningful, but the causes of ulterior motives can also be complex in the business sense. Attackers may have been bought by another big company, affecting their own company's reputation or providing access to certain assets before a deal is made, which could be disastrous for any organization out there.

In the next section, we will see the different types of cyber threat actors and how each of them operates.

Different types of cyber threat actors

A growing number of security breaches make it imperative that we understand the threat actors behind them, what they are capable of, and what motivates them. It is important to remember that threat actors have varying motivations, but they will all have a *purpose* behind them. In order to learn what this means, we must examine how concerned each threat actor is with being identified by the appropriate authorities. It is commonly thought that cyber attacks are done for financial gain, but that is not necessarily the case, especially when we consider hobbyists who are most likely interested in using cyber attacks for personal gain. The different types of cyber threat actors can be divided into the following categories:

- **Hobbyists**: A hobbyist, sometimes referred to as a *script kiddie*, is usually a low-skilled hacker, often acting alone, without many financial resources, and with little or no prior experience. There is usually a motive behind hobbyists to stay on top of their reputations by exploiting vulnerabilities within technical systems; that is to say, hobbyists are *curious* about technology and seek to improve their reputation through technical exploits. How do hobbyists attack computers? There are indeed several low-level attacks that can be used against your system. Defacing a website is like writing graffiti. A DoS is one of the most common threats to application servers and involves throwing so many packets at a target that it is unable to respond. A SQL injection attack involves tricking the database into changing the page URL, exposing more content than was intended to be displayed due to the SQL injection attack. There is a relative lack of concern about attribution from hobbyists because they adopt a gray area in the law, primarily because of crossovers and violations of privacy or integrity, so there isn't much concern.

- **Cybercriminals**: These threat actors usually run cybercrime networks. Do these actors pose a threat? These criminals have a wide range of skill levels. However, it is assumed that most of them possess more advanced skills than hobbyists, and their motivations purely rely on financial gain. Moreover, they also possess more resources than hobbyists, which means they pose a more significant threat to businesses and organizations than hobbyists do. To make money out of your sensitive personal data, they are primarily concerned with fraud, theft, and extortion tactics as their main foci.

- **Hacktivists**: Regarding threat actors, hackers and activists are different in many ways. The term hacktivist combines the words *hack* and *activist*. Hacktivists are groups of hackers who use hacking to carry out a particular ideological, philosophical, or religious objective. They primarily target information exposure, defacing websites, and DoS attacks.

- **Corporate espionage**: This is where a member of staff, with legitimate access to the corporate data, misuses their approved access to exfiltrate corporate data for personal gain, or to provide it to a competitor or an external party.

In the next section, we will look at a generalized impact assessment of the attacks and threats that we've seen so far.

Impacts of cybercrime

Financial losses can reach millions of dollars in cyber incidents, with recurring costs. Organizations are more concerned about what information customers provide them with as awareness of cybercrime increases. As a result of failing to protect your customers' data, you can lose their trust, and the reputation you enjoy can suffer a great deal of damage. Therefore, investors are at risk of further revenue losses.

As a result of sophisticated cybercrimes, including ransomware and DoS attacks, cybercrime can potentially cause substantial damage to organizations. When ransom attacks occur, not only are business operations halted but they are also expensive, requiring hefty ransom payments at the end of the day.

From this point of view, the recent attack on the Colonial Pipeline in the US can be interpreted as a stark example of this situation. The fuel distribution network was shut down immediately after the attack and people expected gasoline shortages as a result, which resulted in chaos. The attack on the Colonial Pipeline also resulted in a direct financial impact on the Colonial Pipeline as a result of preventing further damage to its reputation due to the 4.4-million-dollar loss it sustained.

In a recent report in Australia, it was suggested that over one-third of Australian businesses hit by ransomware attacks paid the ransom, and most organizations didn't even have a policy for dealing with ransomware attacks in place. Several jurisdictions are under pressure to introduce a mandatory reporting requirement for ransom payments to give a true picture of the extent of the problem across the country.

Organizations have a legal obligation to report how they handle and store information related to their customers as a result of the **California Consumer Privacy Act (CCPA)** and the **General Data Protection Regulation (GDPR)**. Organizations struggle to meet deadlines for these requests without adequate processes in place, which places them in a vulnerable position. Aside from this, organizations can also face heavy fines if they fail to comply with regulatory requirements for matters such as the ethical storage of data. Fines of over £150 million were issued under the GDPR in 2020 alone.

Australians amended a similar privacy act in 2017, which requires all Australian government agencies and private companies with an annual revenue of more than $3 million to report cybersecurity breaches to the Australian Information Commissioner within 30 days of the discovery of the security breach. In the face of increased compliance pressures and the surge in cyber attacks, companies must ensure their cybersecurity.

An approach to security that is proactive rather than reactive

Properly implementing rules and procedures is crucial to preparing for cyber threats fully. The IT department, the business strategy team, and employees should also provide feedback on technology use. Moreover, a balance must also be maintained between the need to protect data and the necessity to ensure that the information crucial for business operations can be accessed easily.

The Ninth Annual Cost of Cybercrime report by Accenture and the Ponemon Institute reported that an organization's average cybercrime cost had increased by $1.4 million in 2021 to $13.0 million, and data breaches had increased by 11% .

Cybercriminals increasingly target businesses and their information stores, which is one of the most expensive and fastest-growing segments of cybercrime and the most commonly committed.

Businesses are increasingly willing to use cloud-based services to store identifiable information, which increases the associated risks. Despite this, it is essential to remember that theft is not the only possible goal for criminals; some may attempt to alter or destroy information to foster distrust in the organization or government.

Ransomware and phishing attacks remain the two most common methods of accessing a business's critical systems or networks, followed by social engineering as one of the easiest ways to hack a business.

Therefore, criminals target third and fourth parties, such as IT providers, to access businesses that they collaborate with regularly, which increases third-party risk. This list of trends and developments has only served to heighten the importance of businesses taking cybersecurity seriously and ensuring they pay attention to it.

Summary

In this chapter, we covered the ways attackers work, analyzed their known techniques, and identified the threat actors who make those threats a reality. Lastly, we covered the impact of cybercrime, how it affects business financially, and how attacks damage the reputation of an organization in the long run. Building a way forward toward a future of defense is what this book was written for. It's good to know what's out there and what a blue team will face, but let's not forget zero-day attacks, which a blue team will know nothing about until an incident response plan is initiated.

In the next chapter, we will look at governance techniques and compliance methods and how those can lead to a more secure world.

Governance, Compliance, Regulations, and Best Practices

In this chapter, you will learn what governance is, how to do it correctly, and how to provide visibility to all the stakeholders in the organization. Next, you will learn why it is important to be aware of any external requirements, ensure governance is carried out at the right level, and lastly, what to expect from the major regulations, such as the **General Data Protection Regulation (GDPR)**.

In this chapter, you will learn about the following:

- Definition of stakeholders and their needs
- Building risk indicators
- Compliance needs and the identification of compliance requirements
- Assurance of compliance and the right level of governance

Definition of stakeholders and their needs

One of the most important things to consider when talking about governance is defining who the stakeholders are and what their activity in the organization is. There can be two types of stakeholders:

- Internal stakeholders
- External stakeholders

The governance team should identify internal stakeholders as corporate directors and employees who are involved in the corporate governance process.

External stakeholders, on the other hand, may include creditors, auditors, customers, suppliers, government agencies, and the public at large.

These stakeholders are influential but not directly involved in the process. The key to stakeholder theory is the realization that all stakeholders are involved in the business in some way, with confidence or anticipation that the business will provide the kind of value anticipated or predicted. Among these benefits may be dividends, wages, bonuses, additional orders, new jobs, tax revenue, and so on. A governance team must satisfy the needs of all stakeholders simultaneously. There will be a priority list regarding which stakeholder to satisfy first. However, that list changes according to the corporation's needs and who is involved.

For a drug development corporation, for example, the needs that the business must satisfy will depend on the regions in which the corporation is based – for example, **Food and Drug Administration (FDA)** regulatory authorities, **European Medicines Agency (EMA)** regulatory authorities, **African Vaccine Regulatory Forum (AVAREF)** regulatory authorities, **National Medical Products Administration (NMPA)** for Chinese regulatory authorities, or the **Therapeutic Goods Administration (TGA)** regulatory authorities. If the business is international, then it must satisfy the needs of all these authorities.

The good news is that if one of these authorities approves a type of medicine, then the rest will follow. For example, if the FDA regulatory authorities approve the drug for malaria for distribution in the American region, then their counterparts, the EMA regulatory authorities, will approve it for distribution in the European region. This approval process usually takes years to complete and sometimes, drugs are not distributed or are only distributed due to emergency approval. The emergency approval process is a requirement when there is a great need for any drug or vaccine.

Upon meeting the requirements of these regulatory authorities, the organization can distribute the drugs or vaccines that meet the needs of the community.

In information security, the needs of other authorities such as privacy regulators and external and internal auditors have to be taken into account. If an organization can meet these needs, then it can achieve a good security posture worldwide.

For example, an organization wants to meet the ISO 27001/2 requirements. They have hired an auditor to test whether their systems meet the requirements of the standard. This would require internal stakeholders to take part in the audit. As mentioned in *Chapter 4, Blue Team Operations*, a system comprises not only technology but people as well. The auditor will identify who the stakeholders for this standard are and make sure they meet the requirements of the ISO standard.

In the financial sector, many regulations have to be met if an organization wants to interact financially with many different countries. For example, in the US, a certain framework called the **US Financial Regulatory Framework (USFRF)** helps organizations by ensuring market efficiency and integrity, customer and investor protection, capital formation and access to credit, illicit activity prevention, taxpayer protection, and, finally, financial stability. These regulatory goals are sometimes complementary but conflict with each other in other instances.

For example, without an appropriate level of consumer and investor safeguards, fewer individuals would be willing to participate in financial markets, and competence and capital establishment could suffer. However, at a certain point, too many guarantees and protections for consumers and investors could make credit and capital ridiculously expensive, reducing market competence and capital development. Regulations generally aim to seek a middle line between those two extremes, where the regulatory burden is as low as possible and the regulatory benefits are as large as possible. Therefore, when taking any action, regulators seek a balance between their various goals.

In the privacy sector, the GDPR plays a noteworthy role and makes it clear that all people are equal in their rights. The GDPR is one of the most comprehensive pieces of legislation passed by the EU in recent memory. It was introduced to standardize data protection law across this single sector and give people in a growing digital economy greater control over how their private information is used.

All organizations that process personal data and operate within or sell goods to the EU are impacted by the GDPR. The definition of processing is designed to cover practically every type of data usage and includes gathering, storage, recovery, amendment, and destruction.

The GDPR applies to both data controllers and processors. Data controllers govern the purpose with which and the manner in which data is processed. Data processors are any third party that takes responsibility for data processing on behalf of a controller.

All those regulations require cooperation between the compliance blue team and the law division of an organization. The law division will study the law, and the compliance blue team will figure out how to monitor the policies and procedures that arise from the law.

In the next section, we will talk about how to build risk indicators as part of the processes of the blue team.

Building risk indicators

Look back at *Chapter 2, Managing a Defense Security Team*, for a definition of **Key Risk Indicators** (**KRIs**).

When developing a KRI, knowledge of an organization and how it operates – as well as knowledge of the possible risks, threats, and vulnerabilities it faces – are key starting points. Without understanding an organization, it is tough to identify where it may be at risk.

Afterward, key operational aspects of the organization are mapped to internal and external risks to identify how those key aspects could be disrupted.

Therefore, features of a good and measurable KRI include the following:

- Details on who is affected, which processes, and which technologies are at risk; where the risk takes place (so what facilities are affected); and other organizational characteristics most important to the organization's continued operation and success

- Recognition of the risks, threats, and vulnerabilities the organization faces, based on their probability of occurring, their operational and financial impact on the organization, and its ability to mitigate the event

- Rating the business characteristics in terms of their exigency to an organization
- Rating of risks, threats, and vulnerabilities in terms of their prospective loss to the organization
- Linking the key business characteristics to the most notable risks to recognize the issues of greatest urgency to an organization
- Metrics to recognize when and how a recognized risk becomes a significant threat to important elements of an organization
- The continuous process of reviewing KRIs and their metrics to recognize any changes that may require management review and feasible action
- Approval of those KRIs by higher-ranking management

It is recommended that KRIs are found and built out by risk management blue team personnel and communicated with the board and the rest of the personnel. The blue team should always be vigilant and able to research new KRIs.

Examples of KRIs are shown in the following table, along with which team each KRI affects and how it is defined:

Type	Affected Teams	Definitions
Mean Time to Repair (MTTR)	Service desk team	The average time (measured in hours) required to restore full functionality of a system or application after a failure (e.g., a service outage), measured from when the failure occurs to when the repair is completed and extended to all required sites (servers, devices, workstations, etc.)
System availability	Service desk team	The time interval (measured in minutes) during which *all* systems are connected and operational for all authorized users, divided by the total time interval during which these systems are scheduled to be available for use during the same period, given as a percentage
Employee Satisfaction Metric (ESM)	Human resources team	The amount of employee satisfaction (measured as a percentage) with the organization, usually measured quarterly

Type	Affected Teams	Definitions
Employee Engagement Metric (EEM)	Human resources team	The amount of employee engagement (measured as a percentage) with various activities within an organization (i.e., training camps, artistic demos, etc.)
Value at Risk Metric (VaRM)	Financial and compliance/risk management team	A risk measure expressed as a number, calculating how much the organization will lose if a given event occurs
Regulatory changes indicator	Compliance/risk management team	A risk measure calculating how many times a regulation has changed in a specific country and how the organization will respond to these changes

Table 6.1 – KRI examples

The blue team should refer to each of these indicators and make sure they are included in quarterly reports. There are many more indicators that the blue team can identify according to each organization's area of expertise. For example, in the pharmaceutical industry, certain requirements and regulations need to be measured and identified by every pharmaceutical organization's blue team.

In the next section, we will delve deeper into compliance needs and how the blue team handles those needs.

Compliance needs and the identification of compliance requirements

Regarding compliance, we mentioned a few bits in the last section but now it is time to put those things into action. The question to answer is how the organization achieves compliance with regulations and standards in various sectors.

Standards are identified as international regulations that require compliance. If an organization does not comply with these standards, then it cannot commit to international commerce or work in a country other than the country in which it is based. Regarding the country in which the company is based, the regulatory authorities of that country are the ones that should audit the company and see that they comply with their regulations.

As mentioned in other chapters of this book, NIST is one of these standards, which has many parts that the organization should consider complying with. NIST is primarily concerned with information security, security in the general sense, and cybersecurity.

NIST has three general parts:

- NIST **Cyber Security Framework (CSF)**
- NIST 800-53
- NIST 800-171

The CSF was released in February 2014 as voluntary guidance based on existing standards and practices for critical infrastructure for companies to improve their security risk management. It is considered the gold standard for building cybersecurity programs, along with being a scalable and customizable approach that can work in organizations of any size across many industries.

The framework covers 23 categories and 108 security controls organizing cybersecurity into 5 core functions that we will see mentioned in many parts of this book:

- **Identify** – Involves assessing and uncovering cybersecurity risks to systems, assets, data, and capabilities. We've covered this part in *Chapter 3, Risk Assessment*.
- **Protect** – Involves developing and implementing safeguards and controls to ensure the delivery of critical infrastructure needs. This is covered in *Chapter 8, Detective Controls*.
- **Detect** – Involves developing activities and controls to monitor and detect cybersecurity events. This is covered in *Chapter 9, Cyber Threat Intelligence*.
- **Respond** – This is considered a red team's bread and butter, but a blue team member should also be able to control and mitigate cybersecurity incidents.
- **Recovery** – Involves developing and implementing processes to restore capabilities when and if an incident/disaster occurs. This is covered in *Chapter 10, Incident Response and Recovery*.

The CSF can help organizations address key security challenges in their technological environments, such as the following:

- Uncovering hidden risks and vulnerabilities
- Leveraging the right tools and resources to address risks
- Prioritizing risks to focus on critical threats
- Understanding which assets need protection

Having identified our assets and what we need to protect by placing controls and detection mechanisms, along with mitigation and other response practices, including recovery measures, we need to create a target profile. A target profile includes what assets we have as an asset inventory without the controls. In a current profile, the organization can include the controls in the asset inventory. Last but not least, the organization must carry out a gap analysis of where it wants to get to in terms of cybersecurity. This last aspect can help an organization reach new heights without bringing the current processes to a halt because this is a continuous process that can change every day. Once the organization has finished with this, NIST includes four implementation tiers to indicate where an organization is within the cybersecurity ecosystem:

- Partial

- Risk-informed

- Repeatable

- Adaptive

In the first tier are organizations that are new to cybersecurity, have not implemented any controls or processes, and are waiting for an attack to happen before implementing any protection or change to their procedures.

Next are the risk-informed organizations that understand what is going on out there and are paying attention, but have no policies in place for when the attacks happen.

The third tier pertains to organizations that have procedures and policies in place but not enough tools and controls to protect their landscape – they can defer and detect what is happening but cannot mitigate those attacks on a real-time basis, as situations change or prevent this from happening.

Lastly, the most evolved of the tiers are adaptive organizations that have implemented enough controls and procedures to deal with attacks on a real-time basis, isolate these attacks, and make sure they have the minimum impact on the organization's assets.

The question stands: how does anyone comply with these requirements? The **Chief Information Security Officer (CISO)** must assemble a team, conduct a risk assessment and gap analysis, apply security controls, create documentation, and undertake staff awareness training. And that's before they even get into internal audits and certification audits. To make matters more complicated, once the organization has been certified by NIST, it must maintain its compliance status by applying new controls or improving its risk assessment capabilities.

An organisation's NIST certification is not strictly necessary as long as the business doesn't have anything to do with the US government. However, if it does, then the organization needs to be audited to determine whether it complies with the standards. However, a good organization will have auditors internal or external who will check for any needed changes to procedures or controls.

There are seven steps to complete regularly:

1. Continually test and review.
2. Keep documentation up to date.
3. Perform internal audits.
4. Keep senior management informed.
5. Establish a regular management review process.
6. Stay on top of corrective actions.
7. Promote ongoing information security staff awareness.

Step 1 is a general requirement for information security – risks will always be visible in any organization, and they are always evolving, so organizations need to keep up and test and review, test and review, test and review. Why three times? The first time involves the identification of risks, which are tested and reviewed. The second time, those same risks are tested and reviewed, and here we can identify what we have covered with the controls. The third is for any new risks that might come up – and the circle repeats itself.

Step 2 relates to all the documentation the blue team constantly maintains. This documentation, whether it is security-related or not, should be kept up to date. It is good to archive data because when reviews are taken care of or when an organization wants to update data they can revise those old documents and renew them.

Step 3, an internal audit, provides a comprehensive review of the effectiveness of the organization's **Information Security Management System (ISMS)**. Alongside a risk assessment and a documentation review, this will help assess the status of an organization's NIST compliance. Internal audits are part of the initial process, so the framework should already be on hand and be able to be repeated as part of compliance maintenance. If an organization doesn't have the framework ready, then following the controls mentioned previously, they can apply a general framework.

These steps can be accomplished according to the organization's specialties in different orders. There is no order regarding the controls – from experience, applying a framework may not work for our organization, so we must create our own way to make things easier and faster. However, usually, organizations can take up to 1 year to be ready to be certified for ISO.

Step 4 is really important, especially for big organizations where not everyone knows what each team is doing. Informing senior management and the board generally should be the CISO's job but when teams make changes to their procedures regularly, they should refer to senior managers who then can talk with the CISO or any other member of the executive board. Generally, **Leadership Teams (LTs)** are a great way to involve all managers in discussions when changes and measures are taken.

As mentioned in *step 4*, *step 5* must have an established procedure such as LTs, where managers are reviewed on the work they complete every semester, or every year – that timeline isn't set in stone.

Step 6 relates to making sure that vulnerabilities are covered, whether in procedural matters or applications. Organizations have to stay on top of these processes and make sure that even a minor tweak to a procedure or an application is documented. This can keep the organization in shape so that if any zero-day attacks happen, they can be handled immediately.

Finally, *step 7* is a constant battle and an everyday reminder that every blue team member should be aware of and should ensure that the rest of the organization is aware of the dangers and threats that are out there.

And that is it for compliance. In the next section, we will focus on ensuring that organizations are compliant and have the right level of governance in place to defend themselves against threats.

Assurance of compliance and the right level of governance

So far, we have delved into how an organization can meet the needs of the many people that are either involved with the organization or surround the organization.

Let's put things into perspective – governance is the framework of authority and accountability that defines and controls the outputs, outcomes, and benefits from projects, programs, and portfolios. Governance is the mechanism whereby the investing organization exerts financial and technical control over the deployment of the work and the realization of the value it creates every day.

Assurance is the process of providing confidence to stakeholders that projects, programs, and portfolios will achieve their objectives to the end of beneficial change. Assurance can also be applied to compliance, where it can make sure that the organization has all the requirements to meet the regulations and standards that apply to it. As we can see from *Figure 6.1*, **Assurance** and **Governance** relate to **Risk**:

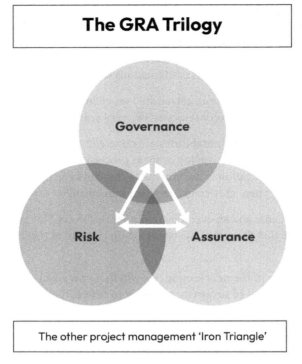

Figure 6.1 – Governance, Risk, and Assurance (GRA) approach

All these aspects are interconnected and need not be taken separately, which is why the people working on each of these parts of the process should communicate and make sure that they all meet each other's requirements. From this, it's clear that governance relies on risk management and assurance, but there are risks that governance arrangements may not work properly. We also need assurance that risk management is good enough. Therefore, we immediately start to see mutual dependence, with the effectiveness of one being reliant on the other two. Governance, risk management, and assurance must all be strong. A weakness in one weakens the effectiveness of the other two.

Risk management is simply another element of project management. Therefore, just like any other process and control, it needs to be assured in project reviews and audits. However, there is never much point in spending much time and effort in assuring things that are low-risk. One of the principles of good assurance is that it is risk-based – meaning evaluating what type of threats the organization faces and how great the threat of each risk is for the organization at that specific period can help assure all stakeholders and thus achieve good governance.

For example, an organization wants to achieve governance in terms of the principle of least privilege but some users have extra privileges and are not denoted as privileged users as they are supposed to be, which is the risk factor. In order to achieve assurance with the stakeholders, the blue team must find those privileged users, document them, and make sure they apply the principle of least privilege, meaning whatever access is not needed business-wise should be revoked.

Another example would be changing the work policies of an organization without applying governance and risk management. This could lead to limited assurance for the stakeholders of the organization, and no governance. This is the risk factor. The business can achieve governance if it applies policy changes, makes sure it meets the requirements of its stakeholders, and meets the regulatory requirements of the country in which it is currently working. Those are the assurance requirements. After all those requirements are met successfully, the organization can achieve governance.

As is evident, compliance is a part of assurance, so in order to achieve governance, compliance is a prime factor. The organization needs to meet the requirements of auditors, regulatory authorities, and so on. Those are some of the stakeholders.

If, for example, it is considered that something in the organization is not compliant, then the blue team should be able to detect that gap in compliance and make sure it is covered so that it can achieve the needed assurance and governance, thus fulfilling the risk management needs of the organization.

Summary

In this chapter, we analyzed stakeholder theory, understood compliance needs using the NIST standard, saw the need for governance, and revealed the relationships between risk management, assurance, governance, and compliance, and where each of these should stand within an organization. This can help you deal with regulatory authorities better, take a strong stance on compliance with auditors, and thus make sure that you meet your goals every year and are not troubled by different regulations or standards, using governance, risk management, and assurance to put them in their rightful place.

In the next chapter, we will be delving into detective controls and how they can detect the threats that concern the blue team.

Part 2:
Controlling the Fray

In this part of the book, we will look into the controls that can be implemented by the blue team to protect assets in its organization. Moreover, we will also review how these controls can be augmented by Cyber Threat Intelligence, and how the blue team could improve the efficacy of its controls.

This part of the book comprises the following chapters:

- *Chapter 7, Preventive Controls*
- *Chapter 8, Detective Controls*
- *Chapter 9, Cyber Threat Intelligence*
- *Chapter 10, Incident Response and Recovery*
- *Chapter 11, Prioritizing and Implementing a Blue Team Strategy*

What are security controls?

Security controls play an important role in shaping the actions taken by cybersecurity professionals to protect an organization. A lack of proper security puts the integrity of content, availability of systems, and confidentiality of data at risk. There are three main types of IT security controls: **technical**, **administrative**, and **physical**.

Security controls are mainly countermeasures to reduce the likelihood that a threat may exploit a specific vulnerability. This act of reducing risks is called risk mitigation. Though it is improbable to stop all threats, all the time, it is still essential to decrease the likelihood of exploitation. This has been explained in detail in *Chapter 3, Risk Assessment*.

The primary goal when implementing security controls is to prevent and reduce the impact of security threats. Therefore, the effectiveness of the security controls depends on the right choice of control according to the risk assessment. The type of controls or level of controls needed for each asset in an organization will depend on the risk assessments performed by the blue team. Hence, not every asset will require the same type of control. Depending on the asset and its risk profile, some assets may need different controls to ensure proper security.

There are five common types of controls, based on their functionality. These controls work at different levels and help protect the required assets of an organization:

- Preventive controls
- Detective controls
- Compensating controls
- Deterrent controls
- Corrective controls

Next, we will briefly look at each of these different types of controls.

Preventive controls

Preventive controls address weaknesses in the information system of the organization, which are identified by the risk management team. They are put into place to avoid an incident occurring, or to reduce its impact in case of any occurrences. They attempt to block unauthorized attempts to access the system, or any data. From a blue team's perspective, preventive controls could include technical measures such as firewalls, endpoint protection, operating system patching and hardening, and email or internet filtration, among many others.

Preventive controls could be process based as well, where the blue team would try and ensure the organization's tech assets are working in a defined way and governed properly. Controls may also include security awareness training for all the staff in the organization and possibly more focused training for senior executives. This helps reduce human error and recognizes that humans will always be the weakest link in security.

The core concept of preventive controls is the effort to avoid incidents and reduce or avoid the need to get into incident management, response, and recovery efforts.

Detective controls

These controls provide visibility into security breaches or any malicious or suspicious activity. Detective security controls function during the progression as well as after the occurrence of the activity. The blue team is responsible for defining the triggers and thresholds of the activities, and alerts are then sent to concerned individuals for action, at the time of detection.

Preventive controls cannot be designed to prevent the occurrence of a threat. Hence, these act as a retrospective check to look for any threats that were not proactively blocked by other controls.

Detective controls use physical, administrative, and technical methods. Physical controls include video surveillance and motion detection, such as activating alarms during the opening of doors without authorization. Examples of technical controls include log monitoring, SIEM, security audits, and the implementation of an intrusion detection system. Lastly, administrative controls include conducting internal audits and finding that there are excess access rights.

As a simple analogy, at an airport, preventive controls are put in place via security guards and immigration counters, to stop any unauthorized person from boarding a flight. On top of that, they may have CCTV cameras as detective controls, which are configured to record and log footage, which can be reviewed as and when needed. Such recording helps the security team look for any threats that may have been able to evade the preventive controls.

Deterrent controls

These controls are implemented by organizations to discourage threat actors from attacking their systems. These could be put up as the first line of defense.

Examples of deterrent controls include security guards, fences, locks, cameras, and alarms. For instance, if an intruder notices that the asset is protected by a surveillance camera, it may deter them from attacking or stealing it. On the technical side, such controls include well-documented policies and regular audits being run in an organization by an independent consultant.

Compensating controls

Compensating controls are alternate solutions to security requirements that are not available for implementation. They are typically workarounds, to compensate for the lack of a primary control. There are two criteria that a compensating control needs to meet:

- They should meet the intention of a primary control
- They should have a similar level of affirmation

For instance, let's assume an organization is planning to use smart cards for strong authentication; however, it faces challenges in procuring the hardware or dispatching it to all its staff. In this case, it may alternatively use software tokens and time-based OTPs as authentication factors. The blue team should run an assessment on these controls to ensure they adequately compensate for the lack of smart cards.

Corrective controls

As the name suggests, corrective controls are implemented to bring the system and resources to their original state if a security incident happens. Corrective controls also include restoration after the incident or recovering the asset to the same secure state as before.

Backups might be the simplest example of a corrective control. If an organization suffers from a ransomware attack, then restoring from backups may possibly be the only alternative it has to recover its lost data.

Corrective controls are useful for an organization to ensure the business does not suffer any catastrophic impact due to a cyber attack. In the event preventive controls are breached, a corrective control may act as a way for the organization to recover and heal from the attack:

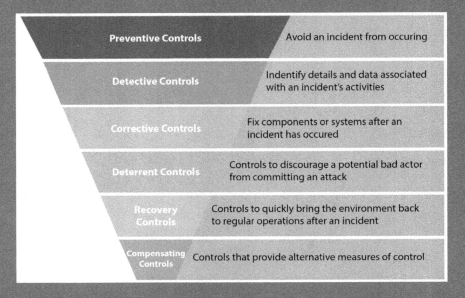

Preventive Controls	Avoid an incident from occuring
Detective Controls	Indentify details and data associated with an incident's activities
Corrective Controls	Fix components or systems after an incident has occured
Deterrent Controls	Controls to discourage a potential bad actor from committing an attack
Recovery Controls	Controls to quickly bring the environment back to regular operations after an incident
Compensating Controls	Controls that provide alternative measures of control

Figure 1 – Types of controls

To summarize, each organization will need a mixture of the five types of controls mentioned. A blue team should run a proper risk assessment to determine the right level of control for its organization. It is also important to note that one single control could work as both preventive and detective measures. For instance, if you look at security guards or surveillance cameras, they serve the purpose of preventive as well as detective control. Corrective controls help improve the efficacy of controls and help an organization get the most value from its investments.

Defense-in-depth

At this stage, it is important to introduce the concept of **Defense-in-Depth** or **DiD**, which originally was developed by the **National Security Agency** (**NSA**) and the military to layer cyber defenses to better protect critical national technical infrastructure.

In a nutshell, this methodology helps the blue team design its security controls in a way that there is redundancy baked in, at every level. If one control were to fail or be compromised, the intention is to have at least one other control be able to defend the organization.

The layered approach of DiD is applicable to all levels of IT systems. The concept is applicable to building the security blueprint of a single endpoint device, as well as an entire organization with tens of thousands of assets. No organization can be adequately protected with a single layer of security. Hence, different defense controls work together to close any potential vulnerability. The main elements of classic DiD security include the following:

- Perimeter security
- Network security
- Endpoint security
- Application security
- Data security

The details of the DiD model are outside the scope of this book. However, it is an extremely important principle to understand for any blue team. In this part of the book, we will deep dive into the preventive and detective controls that any organization should look into. The other controls are equally important for the organization, but we will focus on where a typical blue team's efforts are spent. Hence, the focus is on these two control types. Moreover, we will also deep dive into cyber threat intelligence, as well as incident response and recovery capabilities.

7
Preventive Controls

Now that we understand what security controls are, we will deep dive into preventive controls. We will understand how these are crucial to any organization and why every blue team must plan and implement these controls per the needs of their respective organizations.

The following topics will be discussed in detail in this chapter:

- What are preventive controls?
- Types of preventive controls
- Layers of preventive controls

What are preventive controls?

A preventive control, or preventative control, is a safeguard that is put in place with the intention of preventing a security incident. The purpose of taking preventative measures is to eliminate or reduce as much as possible the potential of a breach, which will inevitably lead to an impact on the organization and the business.

A preventive control makes an effort to thwart any unauthorized efforts to make a change or access a system. This helps preserve the confidentiality, integrity, and availability of IT assets. Hence, a blue team needs to deploy the appropriate level of controls and safeguards, per the risk assessment of each IT asset in its scope.

Moreover, any controls in the organization will need to be regularly monitored and updated to ensure they are running at optimal levels and that the changing needs of the business, the industry, and the threat landscape are taken into due consideration.

Benefits

As the name suggests, preventive controls are placed to avoid any security incidents. They stop any outages or incidents by blocking unauthorized attempts from accessing systems or data. Hence, these are absolutely a must-have for each organization. The benefits of prevention are always going to far exceed the troubles and costs that come with any incident response.

That said, it is equally important for the blue team to be pragmatic in its approach and to select the controls that are the best fit for its own environment and business. As a rule of thumb, the cost of implementing and maintaining a control cannot exceed the value of the asset being protected.

In real life, this is easier said than done. Calculating the value of a business asset is not trivial. An organization's intellectual property, customer data, and employee data all have value, both subjective and objective. This should be kept in mind by the blue team. A security incident could bring in bad publicity and hence impact the goodwill of the organization, but that could be hard to quantify.

Hence, in real life, it is common for a blue team to not put a quantifiable dollar value on each asset. The team may very well put a subjective (high, medium, low, or similar) rating to each asset. This is usually done to help understand the sensitivity level of an asset, from the business perspective.

In a nutshell, note the following:

- Preventive controls help to avoid any security attack before it can cause damage
- Preventive controls save precious data and information as well as the resources put in place in the future to perform corrective measures in case of a security breach
- These controls are cost-effective at times—for instance, simple mechanisms such as passwords prevent any unauthorized users from entering the system

Now that we understand what preventive controls are, let's understand the different types of these controls.

Types of preventive controls

Typically, there are three types of controls that a blue team could consider. Each type of control should be evaluated by the blue team, and assessed per its risk ratings, to select the appropriate level of control. It is also important to note that in real life, one control may span across multiple or even all of these three categories. Hence, blue teams should invest their time in selecting what is the best fit for their setup.

Administrative

Administrative controls are also referred to as *soft controls*. Their focus tends to be from a process or an administration perspective, hence they tend to be harder to enforce. The following are some examples of preventative administrative controls:

- Guidelines and regulations
- Procedures for boarding and recruiting new employees
- Checking references and doing background checks
- Procedures for offboarding and terminating employees

- Training to raise knowledge of security concerns
- The categorization and labeling of data

Physical

Physical controls are the measures and methods that are kept in place to secure a company's premises, workers, and other tangible assets. These include enforcing controls that block cyber-attacks in the real world and hence stop an attacker from being able to physically reach IT assets. They address both the detection of trespassers and the implementation of countermeasures in reaction to risks that have been recognized.

These controls are most commonly used to refer to the process of preventing individuals, whether they be external actors or possible internal threats, from gaining access to locations or assets that they are not authorized to use. This could include preventing individuals from entering a building or a piece of property that they are not authorized to enter. It could mean preventing members of the general public from entering the office headquarters, preventing on-site third parties from entering locations where sensitive work is performed, or preventing your employees from entering mission-critical facilities, such as server rooms. All of these measures are necessary to protect sensitive information.

The following are some examples of preventive physical controls:

- Keycards, biometric IDs, and ID badges
- Fences, mantraps, and locks
- Security guards and guard dogs

Technical/logical

Technical controls, which are also referred to as logical controls on occasion, are controls and techniques—such as software and hardware components—that are set in place to regulate the accessibility of systems. These controls stop logical access to IT assets and use technical controls to prevent any misuse of systems. The following are some examples of preventive technical controls:

- Passwords and biometrics
- Encryption
- Secure communications, least privilege concept, and access control lists
- User interfaces with limited functionality
- Software designed to combat malware
- Firewalls
- Code inspection and analysis using a static environment
- **Intrusion Prevention Systems (IPS)**

Now that we understand the three types of preventive controls, we will deep dive into some of the controls that are usually applicable across various organizations that blue teams should consider. Do note that this list is by no means comprehensive. Since each organization is unique, it is important for the blue team to study its respective business and conduct proper risk assessments before embarking on selecting and investing in the appropriate controls.

Layers of preventive controls

In the introduction to *Part 2, Controlling the Fay*, of the book, we learned about a **Defense-in-Depth (DiD)** approach. This is a good mindset to adopt, and hence this methodology mentions a few layers where controls should be applied. A blue team should look into each of these layers to see which ones are applicable to it, and where its organization's IT assets are located. We will review each of these layers, with some examples of preventive controls that should be considered. You can see an overview of the layers in the following diagram:

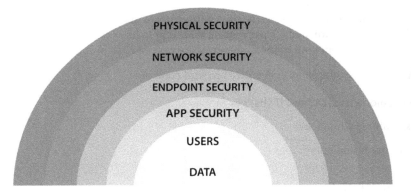

Figure 7.1 – Layers of controls

Next, we will review each of these layers and understand what a blue team should consider when defining controls for its organization.

Policy control

Policy management is required to ensure there is one single strategy and one consistent approach enforced across the organization. Hence, this is usually the first step that is taken before embarking on the journey of getting the relevant controls in place.

Security policies help secure support from senior management and ensure there is a clear message set, right from the top. Secondly, documented processes and procedures ensure there is a consistent approach to security across all assets, in the entire organization. Such documentation not only helps to ensure all staff members across various teams are aligned to the recommendations of the blue team, but also helps ensure that there is a way to ensure continuity of processes and best practices if key staff members were to leave the organization.

These policies must also include technical baselines that should be adopted and configured, across the organization. Examples include baselines for patching, antivirus, hardening standards for operating systems, and any other security tool or control that may be in place in the organization.

Examples of a wide variety of different cybersecurity policies may be found on the **SANS Institute's** website. These SANS templates contain a password protection policy, an email policy, a remote access policy, a wireless communication policy, and many more. These could act as a reference point for the blue team. However, ultimately, every organization is different and hence requires its own set of policies.

It is also important to note that organizations that are regulated may additionally need to check in with their respective regulators for any mandates and/or guidelines that they need to adhere to. These should be baked into the policies as well. These requirements are even more profound in the public sector or organizations in healthcare, financial services, or any other business dealing with personal data. These businesses run the danger of incurring significant fines or in some cases may even lose their business license, in the event the regulator finds their security protocols to be insufficient.

In some cases, organizations need to make their cybersecurity policies public, on their website, to build their public image and establish trust. Consumers, partners, shareholders, and potential employees are all interested in seeing evidence that the company can adequately secure its sensitive data.

As a best practice, these policies, processes, and procedures should be reviewed and updated on a regular basis. This helps ensure the documentation stays up to date and correctly reflects the change in the business and the tech stack being used.

Here are some common policies that organizations require. Once again, the needs of every company are different, and hence this should be looked into by the blue team:

- **Policy for the acceptable use of data systems**: Defines the appropriate usage of tech systems. This could be for internal staff, or for customers/clients. These guidelines serve to safeguard not just the authorized user but also the entire business. Inappropriate use puts the company at risk for a variety of dangers, including the spread of viruses, the compromising of network infrastructure and services, and even legal problems.

- **Policy for Identity and Access Management (IAM)**: This policy's goal is to provide a standard for the establishment, management, usage, and deletion of accounts that are used to ease information access and technological resources inside the organization. The policy should also include checks and controls on how staff could request access to the systems.

- **Policy for the disposal of hardware and electronic media**: This policy extends protection to any company-owned hardware or media that is past the point of use. This policy will create and clarify criteria, methods, and constraints that must be followed to securely dispose of assets and to ensure there is no risk to the organization.

- **Incident management policy**: This policy outlines the requirements for reporting incidents involving the company's information systems and operations, as well as the planned responses to those occurrences. The business gains the ability to respond and recover from an incident in a timely manner post a cyber incident.

- **Password policy**: The use of a username and a password has long been an essential component of security controls for preventing unauthorized access. The policy should include the frequency of password changes, and password complexity requirements. Moreover, there should be details of when and how **Multi-Factor Authentication (MFA)** should be enforced.

As mentioned earlier, these are only some of the many policies to be considered by every blue team, but this should act as a starting point for any good security program.

Perimeter/physical controls

The practice of safeguarding resources and information by putting in place a secure boundary is referred to as perimeter security. In its most fundamental form, perimeter security requires security specialists to create a strategy that relies on a perimeter designed to safeguard machines and ward off threats when they try to enter a secure network. Firewalls and other browser separation systems are examples of the types of technologies that contribute to the perimeter system of a company.

An aspect of perimeter protection that is often the easiest to understand is the barrier that divides your organization's network from the rest of the world. Hence, every network must always have some level of perimeter protection, even if it is entirely detached from the internet. The intention of this layer is to ensure that the network is only limited to authorized users who have a legitimate reason for access. As previously mentioned, a good blue team strategy always involves layers of controls. That stays true for perimeter security as well.

Some common controls that should be considered are set out here:

- In general, the gateway—which could be a router, a reverse proxy, or a load balancer—serves as the first line of defense for the perimeter. The blue team should ensure these assets are configured properly and hardened per defined baselines and policies.

- Next, the network or host firewalls could be configured for a finer level of control. These should help filter malicious traffic packets, and even drop traffic from known threat actors.

- An IPS is a tool that can be a single device or a collection of sensors. This should be configured to identify any suspicious behavior and block that access immediately. It can also be developed to define a typical traffic pattern of the network and look for any anomalies. The systems are configured for signature-based scanning or even behavioral heuristics.

- A proxy service acts as an intermediary between the organization's users and the internet. This can be done by redirecting all outbound internet traffic, via the proxy server. This grants an administrator enhanced control over which users are permitted to connect to the internet and control over which traffic or content is permitted. When a proxy examines the contents of a data packet, it can determine whether the data packet poses a threat by examining information that is included within the packet's header as well as data that is included within the packet's payload.

Similarly, there could be many other controls that could be considered by the blue team. The organization's requirements should be evaluated to see if reverse proxy, mobile security, and other controls need to be put in place.

On the physical layer, it is equally important to protect the office premises, data centers, server rooms, and so on. Physical controls are the implementation of countermeasures to prevent unauthorized access to sensitive material. This would include setting up security guards, CCTV cameras, biometric verification, and so on. Once again, the level of controls would need to be determined per the risk assessment conducted by the blue team.

Network controls

Network control refers to the act of stopping unapproved individuals and devices from joining a private network. This can be accomplished by restricting access to the network. Organizations that occasionally grant access to some users or devices from outside the network to their network could use **Network Access Control** (**NAC**) to guarantee that these devices comply with corporate security compliance standards.

Any network, regardless of how well it is protected, is susceptible to both malicious usage and unintentional harm if it is not adequately secured. It is possible for hackers or disgruntled workers to leak sensitive data, including trade secrets and the personal data of customers.

It is becoming increasingly common for personal devices that are not associated with a company to access the network and assets of the organization. This involves deciding who or what is allowed to access the network and how it is accessed. Network security protects the functionality of the network by ensuring that only authorized users and devices can connect to it and that the devices themselves are free of malware or any other threat a device may introduce to the network.

NAC is one of the elements that contribute to the overall security of a computer network. Although there is a great deal of NAC software available, the tasks are often carried out by a network access server. An effective NAC system limits users' connectivity to only those devices for which authorization has been granted and which satisfy all applicable security requirements. This should include checks to verify that all required security updates and controls have been installed on these devices. The operators of the network are responsible for defining security policies that are used to evaluate if a device or application satisfies the requirements of endpoint security standards and is, as a result, permitted to connect to the network.

The use of MFA could also be considered and either mandated or recommended to the user. This is a much more secure method of authentication. Alternatively, for devices, a digital certificate could also be used to uniquely identify authorized assets.

At the most basic level, IP addresses, hostnames, or combinations of usernames and passwords could be used for authentication. However, this is a very basic check and may be readily defeated by an attacker who could easily spoof an IP or hostname. Passwords too can be compromised via phishing, credential stuffing, or other similar attacks.

Some organizations, especially in highly secure environments, limit the number of untrusted devices or personal devices that can connect to them. In this case, it is possible to use **Virtual Machines** (**VMs**) as part of the organization's data center to benefit from NAC. Hence, the end user would be required to remote into a VM, which in turn could be used to connect to the secure environment.

Once a user has been given permission to access a network, secure NAC can offer additional degrees of protection around specified regions or segments of the network, ensuring that production data or sensitive applications remain safe. It's possible that some of the approaches for regulating access to a network will also contain appropriate security features, such as encryption and increased visibility throughout the network.

In more advanced setups, two layers of controls may be applied at the NAC level:

- **Pre-admission**: This takes place before the user or endpoint device is granted access to the network. This is because it takes place before the user or endpoint device makes a request to enter the network. The pre-admission network control evaluates the access attempt and decides whether or not to grant access based on whether or not the user or device making the request can demonstrate that they are in compliance with the corporate security policy and are authorized to access the network.

- **Post-admission**: A procedure that is carried out inside of the network and given the name *post-admission NAC* occurs whenever a user or device that has previously been granted access to the network attempts to access a different part of the network. In that case, the NAC will be able to censor lateral displacement inside the network and decrease the amount of damage that is caused by a cyber-attack. Users and devices are needed to reauthenticate themselves every time they submit a request to move to a different part of the network. This ensures that only authorized users and devices may access the network.

For a blue team, the NAC may very well be the first layer of control to defend its own organization's network and assets from any potential attacks.

Data security controls

Data security can be defined as the process of safeguarding digitally stored information from unauthorized access, theft, or corruption. Organizations all over the world are today spending heavily to advance their methods of protecting the data of their users as this is an essential asset of every organization. In some cases, data is known as the new digital gold, hence the need for relevant controls is obvious.

Data security is an important aspect of cybersecurity as it protects important information that, if leaked in any possible way, might cause harm to the organization or its users, clients, or customers.

Protecting the important data of companies can save them from huge financial losses. Moreover, it is also good to note that there are obligations from various regulators to safeguard user data from any unauthorized disclosure.

Some common controls around data security are outlined here:

- **Data encryption**: This helps ensure data is protected both at rest on the servers and databases, as well as in motion when being moved between various computing resources or while on the way to the endpoints.

- **Data retention**: This policy mandates the secure deletion of the organization's data when it is no longer needed. Such controls ensure the risk footprint is reduced.

- **Data backups**: This is also a crucial control to ensure that any important data is regularly backed up and is available in the event of any outage or incident. This helps with recovery efforts.

- **Secure disposal**: Before any laptops, hard disks, or any other hardware are disposed of, they should go through a secure data erasure process. This will ensure any organizational data is securely deleted and cannot be recovered by any adversary.

Such controls should be applied not just to the organization's databases, but also to file shares, NAS, or any other data repository. This will help ensure data across all locations is comprehensively and adequately protected.

Application security controls

This control is a set of measures that are followed to keep the code that belongs to the application and the required data and executables/binaries of the application out of reach of any threat actor. Ideally, security controls should be designed, alongside the development of the application, right at the conceptualization phase of the application. This is to keep in mind the use cases of the application, the audience it focuses on, and the design of the application. The security measures of applications should be designed by the blue team.

Next, we will look at some common controls that each blue team should consider:

- **Authentication**: When various sets of controls are being designed for the application, the developers must ensure that the person who gains access to the application is a legitimate user. This may be understood when we enter our password or the user ID in order to enter a particular application. MFA is a more secure choice popular in various applications. It involves the process of authentication at two different levels. Another alternative is to use **Single Sign-On (SSO)**, which helps the user to have a single account and be able to authenticate to various applications using the same set of credentials.

- **Authorization**: After a user is authenticated by the application, there needs to be a validation of the user's access levels and rights. Hence, depending on the user and the system, a user may be granted access to different parts of the application or, alternatively, have different degrees of access. Here, users may be given a **Role-Based Access Control (RBAC)**, which helps define their privileges based on their job responsibilities.

- **Encryption**: Encryption is the process by which the data involved in an application is made unreadable to any unauthorized entity. The application should ensure that the encryption standards defined by the blue team are well enforced and that only the approved cryptographic algorithms are used within the application's code. It is a best practice to use an algorithm that has been well tested by the community and is well trusted by security professionals. Moreover, the developers should also understand the different types of encryptions, not only for protecting the confidentiality of the data but also to ensure data integrity. For this purpose, data hashing algorithms could be enforced.

- **Testing of the application**: Another important step involved that can make your application more secure and prevent hackers from attacking it is by testing it regularly throughout the process of its development, and regularly thereafter. When the application is in its development phase, it goes through various steps of testing. This involves testing the application through various situations in which its data could be mishandled or leaked. Also, simulating tests could be performed for the development team to keep them in a scenario where something goes wrong with the security of the application. This test will help them to act better when a real problem arises. After the software has become productive, regression analysis and tests can ensure that previously tested results are still valid.

These are some of the preventive controls to be considered for application security. Every blue team should remember to enforce these and educate their coding and development teams accordingly. This is especially true for the company's home-grown applications. However, the principle applies to any application that is purchased from external sources as well.

Endpoint security controls

This layer refers to the practice of securing the endpoints or entry points of end-user devices such as desktop computers, laptop computers, and mobile devices in order to prevent them from being exploited by malicious actors and campaigns.

Examples of end systems include desktops and laptops, as well as tablets and mobile devices. Some organizations may even have **Point-Of-Sale (POS)** systems, printers, cameras, and other assets. Endpoint safety is also referred to as endpoint protection. It is important to remember that these endpoints could be employee-owned personal devices, corporate-owned devices, or employee-owned but corporate-managed assets.

More and more organizations have moved on to a hybrid work environment, increasing the use of a **Bring Your Own Device (BYOD)** culture, making these endpoints more attractive to attackers. This has increased access to endpoints for attackers. Unsecured and unmonitored endpoints can access your assets or your organization's confidential information. This enables attackers to expose your data or take it hostage for any unlawful activity.

The quantity and sophistication of cybersecurity threats have been steadily growing, and along with them has come the requirement for endpoint protection solutions that are progressively more advanced. Modern endpoint security systems are designed to recognize, analyze, stop, and contain assaults as rapidly as possible while they are still in progress. As a result, traditional antivirus software has largely been replaced by endpoint security solutions as the go-to option for providing comprehensive defense against the sophisticated malware of today and the rapidly expanding pool of zero-day threats.

In order to ensure that all devices follow a standard security policy, the blue team should start with its baseline policy. Some common controls that should be considered to protect endpoints are set out here:

- Blocking access to unsafe websites

- Classifying data and preventing it from being exposed or leaked

- Host firewalls that prevent malicious activities and block attacks

- Encrypting data on devices, to protect them in case a device were to be lost or stolen

- Predictive and behavioral analytics to ensure cyber threats are blocked

- Regular updates and patches to all applications installed on the device, the firmware, and the operating system

- **Antivirus**: These solutions can scan various files as well as external devices that are connected to an endpoint and identify any threat they may bring

- **Executable control**: This controls the various executables and applications installed on the device. It provides strict instructions to the application as to what it can and cannot do

Once again, every organization is different, and that is why a blue team should ideally take ownership of running a risk assessment to develop the appropriate controls.

User security

It is commonly said that the user is always the weakest link in cybersecurity. This leads to attackers targeting the staff to get a foothold in an organization. Hence, for the blue team user, security is of paramount importance.

User security includes educating and testing employees to help protect your business against cybercrimes, including phishing and other social engineering attacks.

Every staff member, including permanent staff, temporary staff, and even contractors, must be held responsible for security. One mistake by an employee could lead to a cyber incident. Hence, staff need to be made aware of the risks of emails, the internet, and external peripheral devices. Clicking on one link or downloading an unsafe attachment could result in a data breach and damage to your company's reputation.

Just as an example, phishing and spear-phishing are an extremely common form of attack and is particularly dangerous because it relies on human behavior. The intention of this attack is to entice an authorized user of an organization to click on a link or to run an executable on their corporate device. This could be done via emails, SMS, or via a variety of other channels.

To tackle this issue, the blue team must always strategize user-aware training and educate every staff member to recognize, avoid, and report potential threats. Moreover, to test the efficacy of the training and the level of awareness, the blue team may even run regular attack simulations. This is usually done by sending legitimate-looking, but harmless, phishing emails.

User security awareness training provides staff with the information they need to understand the dangers of social engineering and detect potential attacks and helps the blue team better protect its business.

It is also important to note that several laws and regulations mandate each staff member to go through regular cyber training. Hence, the blue team should check on its obligations before designing this control.

Eventually, no control is perfect. Staff must be made aware that no tech control would be able to stop 100% of threats. Hence, the vigilance of staff should always be the key to an organization's defense. Each staff member should go through the requisite training and be asked to report anything suspicious to the blue team.

Summary

In this chapter, we understood what cybersecurity preventive controls are, and why these are essential for any blue team. We also read about the different types of controls and the various layers of controls needed for an effective defense strategy. In the next chapter, we will investigate what detective controls are and why they too are essential for every organization.

8
Detective Controls

In the previous chapter, we understood the importance of preventive controls and what value they bring to an organization. Next, we will understand what detective controls are and how they complement the use of preventive controls in an organization. The objective for every blue team must be to balance the use of the various controls and deploy what is the right fit for its own organization.

The following topics will be discussed in detail in this chapter:

- What are detective controls?
- Types of detective controls
- **Security Operations Center (SOC)**
- Vulnerability testing
- Penetration testing
- Red teams
- Bug bounty
- Source code scanning
- Compliance scanning or hardening scans
- Tools for detective controls

What are detective controls?

The term *detective control* comes from the field of accounting, where one of the mandated controls was to physically audit and count the number of items stored in the warehouse. The intention was to establish whether the actual inventories corresponded with those that were reported in the accounting system.

From a cybersecurity perspective, detective controls are measures put in place to detect and alert on any unauthorized activity on the organization's IT assets. This could help look for any malicious activity in real time or retrospectively.

In a nutshell, detective controls make it feasible to discover cybersecurity incidents in a timely way. Such controls complement preventive controls to ensure no threat can penetrate the defenses and compromise the organization's systems or data. As an analogy, imagine a building that is physically secured by strong biometric locks, with high walls defending the perimeter and patrolled 24x7 by armed security guards. All of these constitute preventive controls. To ensure these controls are not bypassed, the building may also have CCTV cameras to monitor secure locations of the building. These CCTV cameras do not prohibit a threat from breaking in; however, they are instrumental in notifying the security team of any breach.

In the cyber context, the blue team's role is to identify and implement the correct level of preventive controls, to be promptly notified of any cyber incidents.

Types of detective controls

Just as with preventive controls, detective controls too could be classified into the same three broad categories. However, once again, it is important to note that in real-life instances, it is likely that one control may spread across more than one category. Blue teams' mission should be selecting the right controls that are most applicable to their respective organizations:

- **Physical controls** include measures to detect any unauthorized access to a physical building, location, or IT assets. This includes controls such as fences, locks, security guards, security alarms, CCTVs, cameras, motion sensors, and so on. The fundamental intention here is to look for and record any physical breaches where the intruder may have been able to evade the preventive physical controls.

- **Technical/logical controls** include technical measures to protect IT assets. This could include **Intrusion Detection Systems** (**IDS**), access logs, data backups, honeypots, and so on. The value of these controls is to log and detect any unauthorized access to the organization's data or systems. This helps look for any intrusion where the threat actor was able to bypass the preventive technical controls. Controls such as vulnerability assessments and penetration tests also fall in this category.

- **Administrative controls** include mandatory leave and job rotation for staff, to detect any intentional fraud or espionage by internal staff. Most organizations mandate regular independent audits or even external audits to review the security posture of the organization and identify gaps. This helps act as an administrative detective control as well.

For the scope of this book, we will focus on the detective controls that each blue team member needs to consider for their organization. These are some of the controls that are even mandated by various laws and regulations around the world. Hence, it is of utmost importance that the blue team completes its risk assessment and, post that, embarks on a journey to select the right mix of controls it would need to mitigate the identified risks.

Next, we will deep dive into some common detective controls and understand what they are and why the blue team member may need to consider them.

SOC

A SOC is a team of information security professionals that collaborate to monitor, identify, evaluate, and investigate any potential cybersecurity hazards posed to an organization. When looking for signs of a cyber breach, continuous scans of computer networks, websites, computers, end-user devices, and all IT assets are regularly conducted. This helps identify any potential vulnerabilities and, most importantly, any attempts being made to breach the security controls. The SOC is in charge of looking at massive amounts of logs, coming in from every security product and every control deployed by the blue team. It needs to develop rules and identify any exceptions or threats. Next, incident response processes need to be maintained and regularly tested to ensure any breach can be timely managed and contained.

In some organizations, the blue team could consist of a SOC role as well. For some other organizations, it would be imperative for the blue team to work closely with the SOC. Typically, the requirements and monitoring use cases would come in from the blue team and are implemented by the SOC. However, the blue team also needs to monitor the latest threats, trends, and reports from the SOC to understand the changing threat landscape and, moreover, to monitor the organization's known vulnerabilities. This will help the blue team design and upgrade the required security controls, to ensure the organization is well protected from the latest threats.

How does a SOC work?

The SOC is tasked with a number of primary functions, the most important of which are threat monitoring and notification. This includes gathering and processing raw logs from each control deployed within the organization and building the requisite use cases on these logs to analyze and identify any suspicious or harmful behavior. The logs that need to be collected and consolidated should come from a variety of tools and products. The intention should be to cover the full spectrum of the organization, right from endpoints to the core infrastructure, which would include all scoped-in servers and databases as well as network devices. Some of the sources of the logs could be network gateways, IDS, **Intrusion Prevention Systems** (**IPS**), domain controllers, endpoint security tools, host and network firewalls, network proxies, and so on. Essentially, every preventive control should be scoped in, and the logs from various core underlying infrastructures should also be scoped in.

Once done, it would be normal for the log volume to be in tens or even hundreds of millions of line items per day, depending on the size and complexity of the organization. Hence, looking for any potential threat is not trivial and needs to have the right use cases built on top, powered by the right tools.

Some compare this SOC process with looking for a needle in a haystack of technical logs. However, without this process, the value of the security controls may not be fully utilized by the organization. As an analogy, installing CCTV cameras in your office environment is a very good control to detect any intruders. However, the installation of cameras itself is not enough, unless there are automated and manual ways to look at the footage and identify any anomalies. That is the value that a SOC brings.

Any notifications or potential alerts should be sent to members of the SOC team or, possibly, the **Computer Security Incident Response Team** (**CSIRT**) as soon as there is potential evidence of compromise such as discrepancies or abnormal patterns, or other **Indicators of Compromise** (**IoCs**) are found.

In the event of a breach, time is always of the essence. The faster a threat is identified, the faster it can be contained and eradicated. Hence, the blue team should work on building the relevant incident management and response processes to ensure that in the event of a cyber incident, the correct steps are taken to mitigate the impact. These processes will be actioned by the SOC and the CSIRT teams and should be regularly tested and updated to ensure they remain effective.

What are the benefits of a SOC?

At a most fundamental level, each blue team should remember that there is little value in deploying controls unless there is suitable oversight and governance of them. Moreover, any alerts or exceptions raised by these controls must be looked into in a timely manner. Hence, the SOC should be mandated to not only review any alerts raised by any single security product but also to aggregate and correlate the logs from all the products. This is where the blue team must step in and help define the required use cases.

The list of use cases relevant to any organization would vastly differ. However, some basic examples of use cases that the blue team may define for the SOC team are presented here:

- Reviewing system or domain controller logs, for any brute-force or credential-stuffing attempts on any account

- Monitoring the use of privileged accounts, and watching for any unauthorized usage

- Monitoring any alerts generated by the antivirus or antimalware systems

- Monitoring the network and host firewall logs for any suspicious traffic

- Monitoring the health of critical systems, uptime, and so on

- Tracking any unpatched assets in the organization, and tracking any corresponding known vulnerabilities

- Monitoring proxy logs and host-based **Data Loss Prevention** (**DLP**) solutions for any data exfiltration attempts

- Watching for traffic on the network for any port scans, ping sweeps, or connection to any unauthorized ports

- Reviewing the logs of middleware services, **Internet Information Services** (**IIS**), Apache, and so on for any unauthorized attempts

Clearly, this list is not comprehensive. The blue team should consider its entire organization's blueprint, with each asset and control deployed to design the requirements for the SOC.

When set up correctly, the SOC would provide a huge tactical advantage to the blue team. This includes the following:

- Reduced amount of time needed to respond to incidents

- Tracking and analyzing the performance of IT systems, and maintaining a health check

- Reducing the amount of time between the discovery of an incident and the corresponding containment

- Maintaining robust incident response processes and communication channels across the entire organization

- Better management and visibility of cyber incidents, which can be responsibly disclosed to senior executives and also to clients, customers, and regulators

To summarize, the SOC will end up being the core backbone of any blue team. Hence, it is imperative that the controls, scope, and requisite processes for this team are set up in a comprehensive way. We will also look at a few other detective controls that the blue team should consider. However, it is very likely that the SOC acts as a transversal team that receives and processes the outputs of each of these controls.

Vulnerability testing

Conducting a vulnerability assessment is an important step that must be taken in order to find security weaknesses in an application, a network, or even an organization. These assessments can be carried out regularly throughout the firm using software that is designed to do automatic scanning. It is recommended that these scans are done on a regular basis to ensure any newly reported vulnerability is scoped into the latest scan and that the organization stays protected from that threat. The frequency of these scans should be defined by the blue team, keeping in mind the complexity of the organization and the frequency of changes that are deployed.

There are various tools available that offer a breadth of capabilities and features. As with any other product, they should be carefully selected and tested within the organization's test environment before being deployed in the production setup. Some of these products focus on application-level scans, and others may even offer to scan the source code of any in-house developed product to look for vulnerabilities at the code level. The scanning program, which is typically referred to as **Static**, **Dynamic**, or **Interactive Application Vulnerability Scan Tools** (**SAST, DAST, IAST**), performs analyses on the applications. During this examination, the applications are inspected for flaws that might allow for unauthorized access, such as cross-site code, code injection, session hijacking, path traversal, and unprotected server configuration.

Next, the blue team must also consider scanning the IT assets at the operating system level, database level, and even firmware level, as needed. Lastly, network devices such as firewalls, routers, and switches should be scoped into the scans to ensure any known vulnerability is detected in a timely manner.

It is vital to bear in mind that conducting a scan by itself is not enough to ensure the security of an IT asset. The blue team will need to spend time studying the reports and wade out any false positives, or known exceptions, before creating an action plan to either patch the systems or build any mitigating control around it. In some cases, the organizations may even need a vulnerability management policy to document the approved roadmap for each scan, and to manage each reported finding.

It is important to add here that in some complex environments with legacy applications, it may prove to be complex to patch a system in a timely manner. There could be upstream or downstream dependencies to consider, and testing may be required before a system can be permitted to be patched. In such cases, the blue team would need to consider the business criticality of the system, along with the risk of not patching the reported vulnerability and weighing up the corresponding risk. In some cases, it may be fine to delay the patching for a few weeks to enable the dependencies to be managed. In some other cases, the blue team may be asked to build mitigating controls around a known vulnerable system to ensure it is not compromised before the finding is mitigated.

In either case, it is of crucial importance to manage and track each known vulnerability, and at the same time for the SOC to be aware of these vulnerable assets to enable it to monitor them for specific vulnerabilities.

Penetration testing

Penetration testing, or pen-test or ethical hacking, is a form of more intrusive—and more detailed—testing. Both vulnerability testing and penetration testing have the same broad goal, which is to uncover weak points in the security controls of an organization. The key difference is that vulnerability scans are largely automated, and hence produce standard outputs, whereas a pen-test is carried out manually by an individual or group of individuals. The philosophy here is to wear the black hat and get into the mindset of an actual attacker. This team would typically study the IT environment and decide on the best tactic to break into an IT asset, or even the full environment.

Unlike a vulnerability scan, a pen-test would be more subtle. Here, the testers try to take all precautions to defeat the defenses of an organization and to try to breach a system. Hence, such tests will not cause as many alarms to go off at the SOC, as an automated scan. In fact, the primary goal of pen-testers is to evade detection altogether. When done with talented individuals, these scans can highlight issues and vulnerabilities in the network that an automated system is unable to detect. An organization may start with an automated scan but should consider running pen-tests on a periodic basis. This could be done by setting up an in-house team or could even be done via an outsourced professional service.

A pen-tester often is required to make an effort to break into live company networks and simulate a real-life attacker. The testers would use a broad variety of hacking tools, and they may use any one of these tools in order to simulate an attack that would occur in the real world. Depending on the scope of the test, the pen-testers may also engage in social engineering, by trying to persuade the staff of the organization to give up some critical information, or even passwords to the systems. This helps test the cyber-awareness training set up by the blue team as well.

The blue team must ensure that it has the brightest minds working for it for such tests. Finding vulnerabilities by ethical individuals is always better than those being exploited by threat actors for nefarious purposes.

Red teams

Depending on the needs of an organization, the blue team may realize that a regular pen-test may not suffice for the security posture. A dedicated team may need to be set up with the sole purpose of pen-testing every aspect of the organization on a regular basis, if not daily. This helps take the maturity of cyber defense to an entirely different level.

In some cases, an organization's policy may not allow for an external service provider to be able to pen-test an internal network or system. In this case, once again, setting up an internal, dedicated team, in the form of a red team, may be able to help.

A red team analysis is a goal-based activity that requires taking a broad and complete look at the company from the point of view of a possible threat. This is because the activity is designed to simulate an actual attack. The objective of doing a red teaming study is to demonstrate how real-world attackers could chain vulnerabilities, in order to breach a system or exfiltrate sensitive data.

In some cases, red teams may even be asked to test the stability of the systems, by stress testing them to evaluate if the load balancers and the failover work as expected. Hence, in these cases, red teams may try to test the availability of the systems, along with testing the security.

Bug bounty

The one limitation of pen-tests and even red teams is that they are dependent on the skill set of those employed by the blue team. However, there is an opportunity to take the help of the masses and hence have the brightest minds out there to test the security of systems. Such initiatives are called bug bounty programs.

People who take part in such a crowdsourcing project known as a bug bounty system, also known as a vulnerability reward scheme, are eligible to receive incentives for discovering and reporting software flaws. This is because people who take part in these projects are given the opportunity to discover and ethically report software vulnerabilities. Bug reward programs are widely employed as a complementary measure to internal code inspections and penetration testing as a component of an organization's vulnerability management approach.

Such programs are widely popular, especially with tech platforms. These schemes offer monetary awards to security professionals and white hat hackers who come forward and disclose any identified vulnerability or bug in the organization's systems. It is necessary for bug reports to have enough information to enable the blue team to replicate the identified issue and be able to patch it. The blue team should review each reported finding and evaluate the difficulty of executing the identified vulnerability, and the severity of the bug, while deciding on the right level of bounty or compensation to be paid to the security researcher. This will ensure the researchers are well compensated and incentivized to invest their time and effort in testing the systems.

Source code scanning

A source code scan is an automated test of a program's source code that is performed with the intention of locating security weaknesses or vulnerabilities in the code and patching them prior to the application being deployed in production environments. The basic intention here is to strengthen the application code and ensure there are no vulnerabilities at that level.

The main objective is to uncover vulnerabilities, which may include buffer overflows, sloppy use of pointers, and inappropriate use of garbage collection techniques. These are all things that a hacker could be able to take advantage of in order to get access to sensitive information.

There are automated tools in place that can help developers analyze their code and give recommendations based on the findings. The blue team may consider deploying these automated controls to not only educate their coders and to improve the applications, but also to put controls in place to prohibit insecure or untested code from being deployed in production environments.

Compliance scanning or hardening scans

It is also important to note what hardening scans are and what value they could bring to the blue team. Typically, the blue team would be tasked to write and enforce policies around the right level of hardening for every IT asset in the organization. This would include operating systems, network devices, databases, and so on. The blue team may even refer to the **Center for Internet Security** (**CIS**) benchmarks for industry best practices. The CIS is a nonprofit entity whose mission is to *identify, develop, validate, promote, and sustain best practice solutions for cyber defense*.

Once these standards are properly enforced, the blue team should run regular checks to determine compliance with those standards. This helps ensure security hardening is enforced right at the onset of the deployment of a new IT asset. Among other controls, this must include ensuring any default credentials on the IT asset are secured and any unused port on the system is closed.

There are many tools in the industry that can enable the blue team to be able to run both compliance scans and vulnerability scans together. It is critical to note that both of these scans provide value to the organization. Hardening should be a core best practice for the organization, and regular scans ensure those controls remain in place across the life cycle of an asset in the organization.

We mentioned red team, penetration testing, and even bug bounty programs in *Chapter 1, Establishing a Defense Program*. To reiterate, in a typical organization, these teams may not sit under the blue team. These could be part of the red team or, alternatively, be an outsourced service. Nevertheless, the output of these teams is a crucial input into the blue team. The reports produced by these teams, and the technical discoveries made by them, are of paramount importance for the blue team and must be actioned by them in a timely manner.

Tools for detective controls

Now that we understand the scope of a typical SOC and the vast amount of data it would need to consume to identify any threats, the next logical question is: *Which tools are available that can help them achieve that?* Needless to say, doing this manually is going to be next to impossible. Hence, it is essential for the blue team to get an overview of the types of tools available that could assist it in setting up its SOC. For the scope of this book, we will review a few types of tools available, and as previously mentioned, each blue team must run its own risk assessments and evaluation to determine what is the best fit for its own organization.

Threat Intelligence Platform (TIP)

A TIP is a type of cybersecurity solution that primarily focuses on the detection, collection, aggregation, organization, and analysis of threat data from the clear web, the deep web, and the black web. A TIP will gather intelligence that can be put into action from a wide variety of sources and then give that intelligence in a number of different formats. It is also used to locate any IoCs or **Indicators of Threats (IoTs)** within the organization.

The blue team should evaluate the tool to see if it can be used to action intelligence coming in from various sources. Most importantly, a good TIP tool not only helps the SOC understand the intelligence but also helps analyze it and identify any indicators within the organization that the team may need to investigate or action. To ensure the tool is comprehensive, the following are some of the sources it should be able to review and absorb intelligence from:

- **Open Source Intelligence (OSINT)** is information about **Threat Intelligence (TI)** that can be acquired through open sources that are available to the general public. These include security announcement lists on both a national and worldwide scale, in addition to reliable security forums.

- **Social Media Intelligence (SOCINT)** is information obtained from various social media sites and platforms.

- **Human Intelligence (HUMINT)** refers to intelligence that is gathered through cultivating human-to-human contacts in crucial areas. It requires conversing with people as opposed to gathering information through automated channels.

- **Dark Web Intelligence (DWI)** refers to the data on potential dangers that can be gleaned from locations on the dark web where cybercriminals congregate to communicate and trade items. Examples of areas where information can be obtained anonymously include black markets, secret chat rooms, dark web forums, and other venues.

A good TIP will help the SOC team better understand and analyze its logs and be able to detect and respond to any potential threats in a timely manner.

Security Orchestration, Automation, and Response (SOAR) tools

A **SOAR** platform is a tool that enables a SOC or even the blue team to not only react to incidents that have been reported but also to proactively eliminate threats at an early stage. The blue team could use this tool to put in automation rules that will help the tool take immediate action in the event of an incident. Moreover, a SOAR tool could even take inputs from the TIP to action any known IoC or IoT. This is accomplished through the utilization of highly developed automation playbooks for problem investigation, analysis, and reaction functions that cover the whole event's lifetime. The primary goal of the blue team here should be to be able to automatically have the tool take some basic actions to stop or contain a threat, before a more detailed and comprehensive incident response process can start.

In a typical organization, the security stack would include a variety of products. This will range from a number of preventive controls that include proxies, firewalls, endpoint controls, and many more. This entire stack may not necessarily be designed to work together and communicate with each other in an actionable way. This is the gap that is filled by SOAR tools and serves as the glue that binds these tools together in order to orchestrate a variety of different security actions. This allows the platform to provide machine-to-machine orchestration capabilities and helps toward the actioning of the ultimate goals of the blue team.

Security Information and Event Management (SIEM) tools

SIEM tools are developed to look for specific patterns or sequences of events. The capability of SIEM tools to apply dynamic correlation rules to a mountain of distinct and unique event log data enables the identification of threats. Correlation rules and statistical algorithms are utilized on the platform in order to derive usable information from various data sources.

This technology can review logs in order to look for indications of malicious activity. Additionally, because it gathers events from all sources across the network, the system can reconstruct the assault chronology, which contributes to determining the nature of the attack as well as the impact it had. This is possible because of the tool's ability to collect data from a vast range of sources.

In order to get the full value from this tool, it is imperative to be able to comprehensively cover the entire organization and the full suite of the security stack and ensure all logs are duly sent to the SIEM for processing. Moreover, the tool may also offer a range of correlation rules, right out of the box, to help the blue team. However, eventually, it is the responsibility of the blue team to ensure the tool is configured to watch for the right use cases.

In many cases, the blue team may have the SIEM's output feed into the SOAR tool. This takes advantage of the capabilities of the SIEM tool to analyze logs and determine threats and the capability of the SOAR tool to trigger the relevant playbook and take automated corrective action. As an example, in case the SIEM determines an attacker using a specific IP address to port scan the organization's web server, then there could be an automated rule in the SOAR to automatically drop all traffic from the given IP address right at the perimeter firewall, thus avoiding any potential breaches.

Digital Forensics (DF) tools

DF, or cyber forensics, includes processes for the blue team that include the collection, retrieval, analysis, review, and storage of digital evidence. Moreover, it is crucial to note that this process must be conducted in a legally acceptable manner to ensure the evidence can be presented in court, in the event an internal investigation leads to filing a civil or criminal suit. Hence, these tools are critical for the blue team's incident response processes.

In the event of a cyber incident, the blue team may need to contain and isolate the infected asset. Once the infection has been contained and isolated, there needs to be a rigorous investigation into what caused the incident, how was the threat actor able to attack the asset, and a thorough check to see if the threat actor was able to exfiltrate any confidential data. To enable this investigation, DF tools come in handy. It is also crucial to note that in the event the incident needs to be reported to the judicial system, there would be a need to preserve the integrity of the digital evidence. Hence, among other things, DF tools can help the blue team make a digital clone of the asset, which could be used for analysis and investigation, without making any changes to the infected asset itself.

> Tip
>
> It may seem daunting for the blue team to evaluate the various products available in the industry, for each of the mentioned categories. It is not trivial to study a massive number of products, to select the right fit for your organization. As a helpful tip, the blue team could review **Gartner's Magic Quadrant**. Gartner runs research on a regular basis to compare vendors based on their standard criteria and methodologies. This provides the blue team Garter's opinions on how they rate a given product, categorized into one of four quadrants: Leaders, Visionaries, Niche Players, and Challengers. The other alternative is to get feedback from other organizations to see what works best for them. Remember—each organization is different, the needs are different, and the threats they face are different. Hence, a thorough evaluation of each product is the key to the success of the blue team.

It is important to remember that no control or tool in this world is going to completely automate and block every type of threat. Hence, it is important for the blue team to augment its tech capabilities with the right processes. This will form the core of the organization's incident response capabilities. We will look at these processes in *Chapter 10, Incident Response and Recovery*. As a rule of thumb, the blue team should try to proactively block as many threats as possible, and on top of that have the capability to detect any threat that may breach the organization's defenses. Post a detection, the organization's incident response processes should be triggered to contain, respond, and recover from the threat. This is the value of detective tools and incident response processes.

Summary

In this chapter, we understood what cyber detective controls are. Moreover, we looked at some of the types of tools that should be considered to be part of the blue team's arsenal. We also looked at how these tools can help automate the detection and response to any potential incident. We also touched upon **Cyber TI (CTI)**. In the next chapter, we will deep dive into this concept and explore what value it can bring to the blue team.

9
Cyber Threat Intelligence

In the previous chapter, we understood how a blue team should ideally go about selecting, deploying, and monitoring security controls in their organization. We also understood how a **Security Operations Center** (**SOC**) team can get overloaded with too many logs to sift through to identify a potential threat. In this chapter, we will look into how CTI can alleviate some of these concerns and help the blue team identify any potential threats in their organization.

The following topics will be discussed in detail in this chapter:

- What is CTI?
- Types of threat intelligence
- Threat intelligence implementation
- Threat hunting
- The MITRE ATT&CK framework

What is CTI?

Cyber Threat Intelligence (CTI) can be defined as data that is collected, processed, and analyzed to understand a threat actor's motives, targets, and attack behaviors. The fundamental objective of CTI is to enable an organization to make faster and better-informed decisions and to help the blue team become proactive in their defense efforts.

A threat intelligence report provides information about the threats that have targeted an organization or may potentially do so. Information like this is used to prepare, prevent, and identify cyber-attacks aimed at stealing valuable information. In many cases, this intelligence may even help organizations become proactive and defend against attacks, even zero-day threats, before they get hit by them.

A company can be brought to its knees by threats that are unknown, and it can be tremendously frightening. Threat intelligence provides in-depth information about specific threats. This includes information about who carries out attacks, their capabilities and motivations, and the applicable indicators. As a result, organizations can make informed decisions regarding how to defend themselves against threats that are applicable to them.

By gaining valuable knowledge about these threats, building effective defense mechanisms, and mitigating risks that could damage a firm's reputation and bottom line, threat intelligence helps augment the capabilities of blue teams.

The quality of CTI

Before we go any further, it is important to understand that intelligence and also CTI will be everywhere. There is no shortage of sources from where this could be sourced. It is easy for a blue team member to get lost in a sea of intelligence. Hence, as a rule of thumb, it is essential to note the three key facets of CTI. This will help the blue team filter through the noise and find what is important for them and their organization:

- **Relevant**: Blue teams should remember that not all intel will be relevant for their organization. There will be cases where a threat actor targets a select few organizations or industries. Also, some attack vectors may not be relevant to the tech stack being used within the organization. Hence, the blue team must filter down to what is relevant and applicable to their own environment.

- **Timely**: Most of the time, the intel will need to be actioned very quickly. The threat indicators may be applicable for a short duration and may change after. The threat actor may be extremely active for a short duration before going dormant. Hence, actioning the intel quickly is of key importance.

- **Accurate**: Lastly, and most importantly, the intel must be accurate. It must be sourced from reliable sources and must not contain any false positives.

These are the core three tenants of good CTI. All of these should be enforced diligently across all blue teams.

It is also imperative to note that CTI can be obtained from various sources, which include paid and free subscriptions. A blue team should review and understand which feed(s) gives their organization the ideal quality of intel that meets their requirements.

Some popular sources to consider are the following:

- **Information Sharing and Analysis Centers (ISACs)**
- **Information Sharing and Analysis Organizations (ISAOs)**
- Vendors – Subscriptions are available, which are either product agnostic or may even be specific to a certain tool/product
- **OSINT** – **Open Source Intelligence** feeds, available freely online
- Government, **Computer Emergency Response Team (CERTs)**, and regulators – Information disseminated by official channels

As always, there is no single magical formula that would be good for everyone. This is where a solid understanding of your own business and a risk assessment will come in useful.

Types of threat intelligence

Now that we understand what CTI is, let's take a look into the different types of intelligence and how each of these could benefit the blue team.

Different types of threat intelligence can broadly be classified into three categories. Each of these categories serves a specific function in collecting and presenting the data and understanding how it relates to ongoing projects within the organization. As we continue on our journey through threat intelligence, let's explore each of the three types:

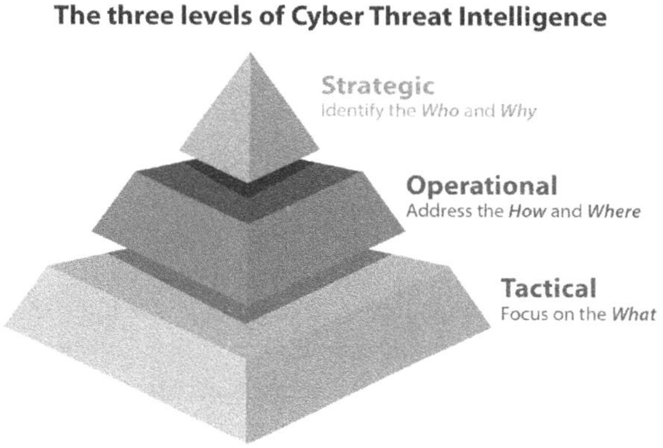

The three levels of Cyber Threat Intelligence

Strategic
Identify the *Who* and *Why*

Operational
Address the *How* and *Where*

Tactical
Focus on the *What*

Figure 9.1 – Levels of CTI

Strategic threat intelligence

A **Strategic Threat Intelligence Feed** (**STIF**), which is often referred to as a high-level intelligence feed, helps in understanding why the threat actors carry out a particular attack in a particular way. Such intel is usually reserved for non-technical audiences, including board members and senior stakeholders. As a result, most of the topics that are covered are those that could have a direct impact on potential business decisions. This intelligence report helps the organization understand the broad trends and motivations that are impacting the threat landscape. The objective here is to bring effective countermeasures.

In contrast to other intelligence categories, strategic threat intelligence may primarily come from open sources or reliable government channels. It is useful to see some examples of these types of media, including local and national news media, white papers and reports, as well as articles and activities available online:

Figure 9.2 – Strategic threat intelligence

This intelligence can be useful for the blue team when looking at the big picture and understanding where the next threat to their organization may come from. It is also important to keep an eye on geopolitical tensions that could cause any adversary to target an organization or sector. For example, before the start of the Russia/Ukraine conflict, there was some good strategic threat intelligence published by various government and private sources advising what threats to expect and what sectors were more prone to cyber warfare. This intel can be more difficult to action, but it is important to provide a broad theme on where the priorities of an organization or a sector should be focused.

Tactical threat intelligence

The objective of tactical threat intelligence is to predict the immediate future of security threats and to assist teams in determining whether or not existing security programs can successfully detect and manage threats. There are various **Indicators of Compromise** (**IoCs**) within a network that can be identified by tactical intelligence, and this allows responders to identify specific threats within that network and eliminate them. An IoC is an example of an actionable threat that a security team needs to know about, such as unusual traffic, red flags raised during logging in, or an increase in file or download requests.

The most basic form of threat intelligence is tactical intelligence, and it is typically automated in order to reduce the time and resources devoted to its generation. Due to the same reason, tactical intelligence generally has a short shelf life since most IoCs become obsolete within a matter of hours after they are created. The type of information presented in this article is aimed at an audience capable of absorbing the information quickly, and it assists security professionals in understanding how their organizations will most likely be targeted due to the latest hacking techniques. Some possible examples here include the IP addresses of the threat actor, the command and control servers, and the hash value of a malicious file.

These indicators help the blue team understand the latest threat vectors and proactively defend themselves. As far as possible, this intelligence should be fed into the SOC automatically, with a direction to proactively block the known attack vectors and also to look for any signs of compromise within an organization.

An example of this type of intel could be the IP address `2.56.59.42`, which is known to be the IP of a command-and-control server. Hence, it would be critical for the blue team to ensure that any communication to this IP address is proactively blocked. Another example here is the domain name `iuqerfsodp9ifjaposdfjhgosurijfaewrwergwea.com`. This domain was used by the WannaCry malware to make a query to check if the machine it was running on was running inside a lab environment.

If logs indicate any communication from an organization's asset to this IP address or URL, it could also indicate a compromise. However, it is trivial for an attacker to change these IoCs, and hence, as mentioned earlier, it is imperative that such intel is actioned as quickly as possible and, if possible, automated to a large degree.

Operational threat intelligence

This type of intelligence is used to answer questions such as *who*, *what*, and *how*. Providing context regarding such factors as the intent, timing, and sophistication of cyber-attacks, it allows security teams to understand the details of specific cyber-attacks.

Analyzing previous and ongoing attacks can provide insight into an organization's adversary's intelligence and capabilities. This information helps defenders identify threats, decipher actor methodology, and act more efficiently when issues arise.

The issue with IoCs is that they are very easy for the attacker to spoof or change. For example, it is trivial for an attacker to use a proxy to change their IP address or to recompile a binary to change the `hash` value. Hence, relying solely on this indicator will not suffice for the blue team. The intention of operational intelligence is to focus on the **Tactics, Techniques, and Procedures** (**TTPs**) of the threat actor. The focus here is to understand the motivation and capabilities of an attacker in order to understand how they could potentially attack you.

This intelligence could be harder to action than IoCs. However, when done correctly, it gives the blue team better security coverage. This is because rather than focusing solely on an indicator that could be changed in real time, the defense is focused on the tactics of the adversary and hence helps make the organization more resilient to attacks.

Some examples of TTPs include malware attack data, cross-industry cybersecurity statistics, incident and attack reports, and other analyses of the threat actor.

Eventually, all three forms of intelligence are going to be of utmost importance to the blue team. In a typical setup, the blue team would be responsible for sifting through such intelligence reports and for ensuring the cyber defenses are set up correctly for each attack vector and for each attacker. In some organizations, the blue team may also be asked by the board to give a briefing on the latest trends and threats that the organization should be safeguarded from. In lots of cases, a threat intelligence report can go a long way in building a case for a new security control or for a new investment.

Threat intelligence implementation

An organization can minimize risk by reducing potential attack surfaces and reducing risk through a CTI life cycle model, which takes raw data and turns it into actionable intelligence from raw data. However, it can be daunting to start the journey of implementing a CTI framework that works for the blue team and their organization.

In order to be effective, cyber threat life cycle models need to take a holistic view and function as an ongoing set of processes that work continuously and loop to identify intelligence gaps and generate new collection requirements, starting the intelligence cycle over again. There are six main steps that make up the threat intelligence life cycle:

1 – Developing a plan

For intelligence to be developed, it is necessary to begin with the right questions in order to develop a process of inquiry. It is better to ask questions that are focused on a specific fact, event, or activity rather than questions that are open-ended. Further, it is also necessary to consider who the target audiences and customers will be, as well as how the information will be consumed. By planning CTI activities in advance, valuable resources are maximized, and results are useful.

2 – Collection

During the planning phase, it is best to consult an array of internal and external sources for raw data. There are a number of data sources available to support the threat intelligence life cycle, including the following:

- Insights curated from threat intelligence sources
- Data collected from the open internet as well as the dark web
- Alerts from internal systems regarding incidents that have taken place
- Sources of information, such as news stories and social media posts
- Identifying malicious IP addresses, domain names, and hashes of files
- Event logs for the network
- Intelligence gathered from open sources

- Keeping track of past incidents and how they were dealt with

- Insights from third parties

3 – Processing

It is important to sort and organize the raw data after it has been collected. As part of the process of creating a robust dataset for analysis, the following conditions must be taken into account:

- Metadata addition is required

- The classification process

- The cleaning process

- Modeling of the data

- Data deduplication

- The enrichment process

- Normalization of data

4 – Analysis

Analyzing raw data is crucial to turning it into actionable information. The purpose of this is to identify suspicious activity and patterns that can lead to security issues. Threat lists and peer-reviewed reports should be used to meet the needs of different audiences. A structured analytical technique is used in the analysis phase of the life cycle of the threat to overcome biases and uncertainties that are inherent in any analysis. As part of the analysis, the following should be considered:

- Correlation between indicators and incidents

- The establishment of relationships

- The structure of data so that it can be indexed and searched

- Creating a visual representation of the data

5 – Dissemination

Dissemination of the findings from the analysis phase should be rapid, and the format should meet the preferences of the consumers. During the planning phase, it should be determined who receives what information and how it will be delivered. There is a growing market for integrating threat data with cybersecurity posture automation platforms for the purpose of delivering a unified, near-real-time cyber risk model that allows for the prioritization of vulnerabilities and their remediation in order to reduce the likelihood and impact of vulnerabilities.

6 – Feedback

During the planning phase, the quality and efficiency of the CTI will have been assessed in order to confirm that the information met the requirements that were established. The collection of feedback is also useful for identifying gaps or mistakes in the threat intelligence life cycle and brings forward more questions or issues to be addressed in the future.

Now that we understand how an organization should go about developing its threat intelligence program, we next need to understand how the blue team should consume this intelligence and hunt for threats in their own organization.

Threat hunting

As the name implies, threat hunting is a practice designed to assist you in finding adversaries within your network. The intention here is to try and look for any compromise proactively rather than waiting for an alert to be raised by the SOC. This helps avoid fire-fighting situations and to try and stop the attackers as early as possible.

It is important to note that a threat hunt is an activity that is quite different from a **Digital Forensics and Incident Response (DFIR)** activity. An imperative aspect of DF/IR methodologies is their ability to evaluate what happened in the event of a data breach after it has already been discovered. When, on the other hand, a threat hunting team engages in the task of threat hunting, they are looking for those attacks that may have already penetrated your defenses but may or may not have led to an impact on your organization. The threat hunting strategy differs from most other forms of security management in that it is a proactive approach that combines the data and capabilities of a security solution with the deep technical and analytical capabilities of an individual or team of people specializing in threat hunting.

The importance of threat hunting

Although automated security tools and analysts in security operations centers should be able to handle most of the threats in your organization, the blue team should still be concerned about advanced threats that are able to evade these tools and controls.

Depending on the organization and the value it has to an adversary, a large portion of the threats may consist of sophisticated threats capable of causing significant damage to a company. The adversary may even have the capability to evade detection from any automated tool or control a blue team may have put in place. There are studies to showcase that an advanced attacker could compromise a network and stay undetected for a period of up to 280 days or potentially even longer. This is where threat hunting comes in and helps the blue team in actively hunting the threat actors. Detecting and effectively resolving threats can reduce the damage that attackers cause by reducing the time between intrusion and discovery.

Data breaches can result in lost revenue, decreased customer loyalty, defections among IT and security personnel, and a poor reputation for the brand. Threat hunting can help mitigate or even counteract these risks. Those organizations that operate with a high level of security maturity, along with a great deal of human expertise, may choose to develop these threat hunting skills on their own, while other organizations, whether large or small, may elect to enlist the services of an external threat hunting firm.

To be most effective, proactive threat hunting must balance people, processes, and technology to identify adversaries faster. Here are some points to keep in mind when a blue team embarks on this journey:

- Establish an executive awareness and involvement program for proactive threat hunting within your organization. Educating organizational leaders on threat hunting could require a cultural shift in your organization, so sharing aggressive offensive techniques such as *hacking back* is off limits.

- Develop an **Incident Response (IR)** as a proactive, rather than a reactive, approach. The first thing you should do is to generate a hypothesis about where potential adversaries may attack your system, application, or logs and then start looking for anomalies within these systems, applications, and logs. Over time, you can refine your approach and build repeatable processes by tracking your progress.

- To effectively counter threats, you need to understand how threat actors operate and what threats the vertical industry sector and your particular organization face. Imagine yourself as a hacker as you identify your company's most valuable and attractive assets to determine how a threat actor might approach these assets and what tactics, techniques, and procedures to follow.

- Taking advantage of existing tools and resources to uncover IoCs, such as **Security Information and Event Management (SIEM)** monitoring, is imperative. With SIEMs such as ours, organizations can easily filter and search data, offering fast searching and easy data filtering. They provide a strong foundation for organizations specializing in security analytics.

- To elevate the outcomes for your threat hunting, there is usually a need to evaluate external support and expertise, which more senior analysts typically conduct. As a result, not all organizations will have the time to devote to threat hunting as well as the dedication, focus, and resources to do it; external organizations and **Managed Security Services Providers (MSSPs)** can support your team or even take over this responsibility with your guidance.

- Lastly, and most importantly, always power your organization's threat hunting missions with the help of CTI. This will ensure the discovery missions are always looking for the latest and greatest threat actors, and the use of TTPs will be the most effective tool in locating the adversaries within your organization.

Now, that said, the next logical question is how could a threat hunter go about looking for a threat? How can a TTP or even IoC be actioned by the blue team or by a threat hunter? Let's look into this next.

Using CTI effectively

A CTI feed forms the backbone of a blue team's cyber defense plan, whether you're looking for a cybersecurity vendor or your employees are going through training. Information and actionable insights provided by threat intelligence services allow organizations to protect themselves against cyber threats. Using CTI can allow businesses to gain access to massive threat databases and improve the effectiveness of their solutions. Ultimately, threat intelligence supports security solutions that determine their effectiveness.

To detect and eliminate cyber threats before they do any damage, a blue team may employ threat hunters, to search for any known IoC in their organizations. Next, let's try and understand what IoCs are.

Indicator of Compromise (IoC)

IOC is a forensic term that refers to evidence that exists on a device that gives a clear indication that a security breach has taken place. An IoC is a system that gathers data after a suspicious event, a security event, or an unexpected network callout. In addition, it is a common practice to periodically check IoC data to detect abnormal activities, vulnerabilities, and other security threats.

A compromise indicator can also serve as a source of information that enables the blue team to take action earlier whenever malignant activity is detected on a network. In this way, it is possible to discontinue such activities before they become actual attacks or compromises, posing a threat to the entire network. Additionally, it is sometimes not easy to detect IoCs because they have varied forms. It could be a piece of log data, metadata, or a complex string of codes. As a result, blue teams must try to make sense of the information being provided to them within the context of the whole system so that deviations can be identified. A further advantage of these tools is that they integrate several indicators to find a correlation between them.

IoC feeds must include the following:

- The behavior of web traffic that is non-human
- URLs that are malicious
- Account activity that seems out of the ordinary
- Attacks related to IP addresses
- Hash values for malware
- Emails with malicious attachments and much more

This will help ensure the threat hunters have the full spectrum of the threats they need to monitor.

Covering the full breadth of threat hunting capability is outside the scope of this book and is a dedicated topic in itself. However, this synopsis should be able to help a blue team understand the importance of this function and how to go about setting one up.

The MITRE ATT&CK framework

MITRE (www.mitre.org) has been a trusted independent adviser since about 1958 and has been working on developing frameworks that help organizations and, more importantly, blue teams to structure their cyber defenses.

They have developed the ATT&CK framework, which stands for **A**dversarial **T**actics, **T**echniques, and **C**ommon **K**nowledge. This framework includes several knowledge points that can be used to build kill chains and threat models that reflect the attackers' behavior and TTPs. It has become a standard to not only determine whether a cybersecurity solution can detect as many TTPs as possible to prove its efficiency but also to define a framework for the sharing of intel between various parties. The framework provides an easy method of sharing intelligence in a uniform language that is both universally accessible and acceptable to everyone.

It is widely acknowledged that the ATT&CK framework serves as an authoritative guide to understanding the tactics and behaviors used by hackers to target organizations. This framework not only removes ambiguity in the terminology and provides a common language that industry professionals can use for discussing and collaborating on this type of threat but also has practical applications that blue teams can use to combat the adversary's methods.

No matter how well-resourced a blue team is, they will not be able to protect themselves from all attack vectors equally. With the ATT&CK framework, teams can generate a blueprint for where they should concentrate their efforts on detecting threats. Many teams may prioritize threats as early as possible in the attack chain. Other teams may select the tactics that are most applicable to their organization or to their own tech stack.

The MITRE ATT&CK Matrix

The MITRE ATT&CK framework currently has three iterations:

- **Enterprise**: Focuses on servers and infrastructure, both on-premises and in the cloud
- **Mobile**: Focuses on the behavior on iOS and Android operating systems
- **ICS**: Focuses on industrial control systems and the actions an adversary may take there

At a fundamental level, the ATT&CK Matrix consists of a set of techniques used by adversaries to accomplish their objectives. These objectives are referred to as *tactics* in the ATT&CK Matrix. Within each tactic, there are the adversary's *techniques*, which describe the actual activity carried out by the adversary. Some techniques have sub-techniques that help clarify the technique in more detail.

Just as one example, the following are the tactics listed for the Enterprise iteration of the ATT&CK framework:

- **Reconnaissance**: This includes a list of techniques that the adversaries use to gather information and plan operations

- **Resource Development**: This includes techniques to support operations, such as setting up command and control infrastructure

- **Initial Access**: This includes techniques used by an adversary while trying to break into a network

- **Execution**: This includes techniques for running malicious code

- **Persistence**: This includes techniques for maintaining a foothold in the network after the initial access

- **Privilege Escalation**: This includes techniques for gaining a higher-level permission

- **Defense Evasion**: This includes techniques for avoiding detection

- **Credential Access**: This includes techniques for stealing account names and passwords

- **Discovery**: Lists techniques for discovery and analyzing the organization

- **Lateral Movement**: Defines techniques for moving within an environment and using one system to access another

- **Collection**: Defines techniques for gathering data of interest per the attacker's goal

- **Command and Control**: Defines techniques for communicating with compromised systems to control them

- **Exfiltration**: Defines techniques for stealing the data and transferring it out of the organization

- **Impact**: Defines techniques for manipulating, interrupting, or destroying systems or data

Similarly, the ATT&CK framework has built iterations for the mobile and ICS environments as well. However, the 14 tactics mentioned previously should give a fair idea of what this framework is about and how it adds value to blue teams.

How to implement the ATT&CK framework

This framework could prove very useful for a blue team embarking on its journey of setting up a CTI program. There are various recommendations for the use of this framework, however, the key best practices are as follows:

1 – Plan a cybersecurity strategy

ATT&CK can help you develop a cybersecurity strategy. Blue teams should plan their controls, building defenses against each tactic on the framework.

2 – Run adversary emulation plans

Using ATT&CK for opponent emulation plans can help improve the performance of a red team. When an organization's red team uses the same framework as the blue team, the synergies between them improve, and they are able to share recommendations between them in a more structured manner.

3 – Identify gaps in defenses

In identifying and addressing gaps in defenses, ATT&CK matrices can assist blue teams in better understanding how a potential cyber-attack might unfold. The ATT&CK documents recommend remediation and compensating controls for the vulnerable assets in the organization.

4 – Integrate threat intelligence

Threat intelligence can be effectively integrated into cyber defense operations with ATT&CK. An implementation plan can be developed to address threats by mapping threats to specific attacker techniques.

Blue teams should remember that this framework will help them not only be able to use and action CTI more effectively internally within their organization but will also help in case they were to share the intel externally with other organizations, industry associations, or with government agencies.

At this stage, it is also important to mention another framework developed by MITRE, which is called **MITRE D3FEND**. This framework is designed for defensive countermeasures to help blue teams plan and tailor their defense for the mentioned MITRE ATT&CK tactics. The D3FEND Matrix includes countermeasures at every stage of an attack, that will help a blue team prevent, mitigate, remediate, and respond to any type of attack.

The D3FEND Matrix is divided into five main defensive tactics: Harden, Detect, Isolate, Deceive, and Evict. Beneath each of those tactics are countermeasure technique categories, including network traffic analysis and user behavior analysis.

This framework, too, is of value for each blue team. This can help ensure a blue team has the right level of control at each layer. For the scope of this chapter, we will not deep dive into this framework, but would still recommend it to blue teams to learn and adopt.

Summary

In this chapter, we have understood what CTI is, and how it adds value to the mission of a blue team. Moreover, we have also looked into the qualities of a good CTI report and, most importantly, how a blue team can sift through the noise to find intel that is most applicable to them. We also looked at some industry-renowned frameworks and the best practices for embarking on the journey to set up a CTI program. In the next chapter, we look at how a blue team should focus on cyber incidents, and plan to respond and recover from them.

Incident Response and Recovery

In previous chapters, we learned how to prevent and detect a threat. Now, it is time to face the threat head-on. We will learn what to do if something happens to the organization and how we recover the business if it's lost due to an attack – that is, by using incident response plans and disaster recovery plans.

In this chapter, you will learn how to make incident response and disaster recovery plans, how to test those plans, and what to do with cyber insurance. This chapter will also cover the NIST: Respond & Recover methodology and explain it thoroughly with examples from incident response teams.

In this chapter, we will cover the following topics:

- Incident response planning
- Testing incident response plans
- Incident response playbooks
- Disaster recovery planning
- Cyber insurance

Incident response planning

In previous chapters of this book, we saw that planning is really important and that prevention is a must to stop incidents from occurring. In this section, we will see what happens when we need to respond to incidents and recover from those incidents.

An incident response plan is really important because without it, the teams working on any project will find themselves in a really hard situation when responding to incidents. For example, a team was building an application and had already launched that web application online. One night, a serious **Distributed Denial-of-Service (DDoS)** attack coincided with a network intrusion and the team had no idea how to respond to that attack. The developers did everything they could to stop the attack from occurring but lost important data that was on the server at the time. This is where Blue teams come in. The Blue Team member who is assigned this case should follow the procedure in place and document any action. Stopping a DDoS attack, along with intrusion attempts, is not that simple. However, following procedure is a must and a plan must have already been written by the CISO or any other team member who handles planning.

The plan should include a purpose, security contacts, data flow, recovery options, incident response training, incident response plan testing, definitions, supporting materials and references, and revision history. Let's look at each in more detail:

1. Firstly, regarding the purpose of the plan, the reason for developing an incident response plan is described. Remember that an incident response plan can be written for any incident because it can refer to various teams within the organization.

2. The second part, security contacts, is the most important, especially if a random member of staff wants to refer to the plan. This can include internal application and business resources responsible for responding, as well as external suppliers that may be involved in an incident. A table of those resources can be provided.

3. The third part requires data flows, which facilitate response activities by providing an understanding of the relationships among information entering/supplied to and produced/delivered from any business process.

4. The fourth part involves the recovery options, which can help the blue team member who is assigned a specific incident case. This can include a description of how the platform can be re-established. This can also include logs that may be helpful to identify or respond to events identified via the *threat analysis* work of the Blue Team.

5. The fifth part contains a description of which logs should show activity on the system, which can be related to the creation or modification of information, or access to that information. This can also include logs that may be helpful to identify or respond to events related to items identified through the *threat analysis* work of the Blue Team.

6. The sixth part consists of the training provided by the organization for responding to incidents. This can include a description of what training is provided, who must take this training, and how often this training is administered within the organization.

7. The seventh part describes how to test the incident response plan, which includes who will initiate exercises to test the plan, how often and when the exercise will be performed, and potential scenarios to be exercised.

The last three parts of the incident response plan include a glossary of terms used in this document, records of supporting materials that can help users if they require more understanding of how this plan became a reality, and a revision history of who revised this procedure and when the last revision was created in the organization.

Having set this plan in place, in the previous situation, where there was no initial plan for DDoS attacks, a blue team member can read the plan and understand who the must contact, as well as make any necessary modifications to the system so that it can continue working without fail.

In this case of DDoS attacks, the Blue Team member should move to the **Web Application Firewall (WAF)** and change the settings of the firewall for DDoS attacks, which will make the firewall stop the big traffic to the website. This may cause some issues with some users, but the firewall can instruct them to wait in a line of other users entering the website at that specific time. A WAF not only helps to protect against DDoS attacks but can also stop most OWASP-related attacks, such as XSS and SQL injections.

Next, the Blue Team member can look at the statistics in the firewall and stop the traffic or enable the DDoS attack mechanism for specific areas only.

Last but not least, the blue team member should write a report that will be sent to the CISO or any other senior leader who is assigned to receive such reports. After that, the senior leader can use such reports to calculate the risk for DDoS attacks yearly or quarterly during the risk assessments conducted by their team.

So far, we have how to set up incident response plans. In the next section, we will test those plans and see how those tests are conducted by senior leadership.

Testing incident response plans

In this section, we will demonstrate how to test the incident response plans we learned how to create in the previous section. NIST has created a revision publication called *IRS Safeguards Technical Assistance Memorandum Incident Response Test and Exercise Guidance*. This section covers *Special Publication 800-84, Guide to Test, Training, and Exercise Programs for IT Plans and Capabilities*, which provides incident response test and exercise guidance and best practices.

This publication defines tests and two types of exercises:

- **Tabletop exercises** are facilitated, discussion-based exercises, where teams can meet and discuss how to resolve an incident if it occurs and have a clear understanding of the roles that each part of the team must play if an incident occurs. In the DDoS scenario, this can be a way to build the incident response plan or to solve any questions about who does what in an incident situation. They can be really useful but not as strenuous as functional exercises.

- **Functional exercises** allow personnel to validate their readiness for emergencies by performing their duties in a simulated environment. Such exercises can strain the limits of what each employee can do based on their role. For example, in a DDoS attack that happens alongside a SQL injection attack scenario, the Blue Team members each have a role to play – one must check the backups, and the other checks the WAF. If all goes well and the backups are recent enough, then none of the data will be lost from the servers and operation can return to normal after the exercise is completed.

- **Tests** are tools that can evaluate whether a team can conduct incident response by using quantifiable metrics to validate the operability of an IT system or system component in an operational environment. A test should be conducted in as close to the same operational environment as possible. In the DDoS example, such tests could include setting up a test server and testing how long it would take until the server fails and then how the Blue Team can restore that server with minimal losses.

NIST also defines the **Incident Response (IR)** life cycle, which should include various stages that are part of their corresponding CSF part:

CSF Part Covered	IR Life Cycle Stage	Summary of Incident Response Activities
Identify	Preparation	• Provide training and awareness for all individuals in the organization so that the users can recognize any anomalous behavior and conduct specific reporting requirements for suspected breaches • Gather contact information for any staff that can handle an incident • Gather hardware and software needed for any technical analysis your incident response team may need to conduct • Perform evaluations, such as tabletop exercises, to test the incident response capability and whether the incident response team can accomplish the goals set in the incident response plan
Detect	Detection and Analysis	• Monitor IT system protection mechanisms and system logs • Investigate any reports conducted by the teams to discover whether they are incident-related
Respond	Containment	• Choose and implement a strategy to prevent further loss based on the level of risk associated with a specific threat • Gather and preserve technical evidence, if applicable
Respond	Eradicate	• Eliminate any components of the incident, such as deleting malicious code or disabling breached accounts, if applicable
Recover	Recovery	• Restore systems via appropriate technical actions, such as restoring the data lost from clean backups, rebuilding the systems from scratch, replacing compromised files with clean versions, installing patches, changing passwords, and tightening network perimeter security

Table 10.1 – Incident response life cycle

According to the NIST publication, there should be material that explains those exercises and their goals. That material is separated into three types:

- **Facilitator guide**: This should include the narrative scenario, a list of questions to guide the exercise, and the incident response plan that is being exercised. This will be used to evaluate any part of the exercise, but it should not include previous solutions that other teams used.

- **Participant guide**: This should include the same essentials as the facilitator guide except for the list of questions.

- **After Action Report** (**AAR**): This provides evaluation criteria based on the exercise's objectives and can be a means to evaluate how well the objectives of the exercise were met. It also identifies areas for improvement and where additional exercises might be necessary.

Let's put the DDoS example into two narrative scenarios that the NIST guide simulates:

Attack Scenario	Tabletop Exercise Objectives
Scenario 1: A website server was attacked last night via a DDoS attack. The logs indicate that packets were received at a very fast rate and consumed the incoming and outgoing bandwidth of the website. **Scenario 2**: A website server has been acting strangely and sending incomplete requests to other website servers inside the organization. This has caused the servers to overflow their concurrent connection pool and has led to a DoS for additional connections from legitimate clients.	• Determine any actions that could help prevent this type of incident (preparation) • Determine any controls in place that could help identify this incident, along with procedures on how to report the incident (detection and analysis) • Determine how to prevent further damage (containment) • Determine how to clean the system (eradication) • Determine how to restore the system securely (recovery)

Table 10.2 – Sample incident response evaluation scenarios

Evaluating the exercise is a critical step to ensuring the success of the incident response program. After the test or exercise has been completed, the participants should conduct a debriefing to discuss observations for things that worked well and things that could be improved. The comments that surface during the debriefing, along with lessons learned documented during the exercise, should be captured in the AAR. The AAR is an important document for the leader who conducts the exercise, helping them understand how an incident response team addresses the issues, the responsibilities of each participant, and any recommendations about improving the method used in the exercise. Don't forget that anything that was not documented never happened.

Having seen how to test the incident response plan, now, we will focus on the incident response playbooks and how to deal with different incidents that we may face.

Incident response playbooks

During the next subsections we will be giving a better understanding of how a Blue Team follows a specific playbook to address the attacks specified in each subsection. Most of the steps are the same but they can follow different ways of determining how an attack is classified. In the following subsections we will be putting forward the following playbooks:

- Ransomware Attacks Playbook
- Data Loss/Theft Attacks Playbook
- Phishing Attacks Playbook

Ransomware attacks Playbook

Nowadays, ransomware is a top threat, and their operations will mostly have similar patterns of attack frameworks, tools, and techniques across victims. They will also have similar operations to other ransomware families such as Ryuk and DoppelPaymer. The same applies to how hackers operate using a wide range of ransomware flavors. In summary, it is a requirement to quickly understand their overall pattern of operation.

The main goal of ransomware is to encrypt all files that it can in an infected system and then demand a ransom to recover the files. However, for example, in Maze, the most important characteristic is the threat that the malware authors give to the victims where, if they do not pay, they will release the information to the internet.

For this threat in this playbook, we can have the following sequence of events:

1. **Preparation**:

   A. Review and practice cyber incident response procedures, integrating both technical and business roles and responsibilities, escalating as necessary to manage major incidents.

   B. Review past and recent cyber incidents and outcomes.

   C. Review threat intelligence data, including ransomware tactics and techniques, to identify threats to the company, brands, and industry, as well as common patterns and emerging security risks and vulnerabilities.

   D. Ensure that the entire incident response team has access to all required documentation and data.

   E. Identify and acquire all necessary forensics tools and equipment.

   F. Define the threat and risk indicators in the framework of the company's **Security Information and Event Management** (**SIEM**) solution.

G. Conduct recurring security awareness campaigns to raise awareness of the cybersecurity risks to which employees are exposed. It is important to keep all teams vigilant to threats and make sure they report all incidents to the Blue Team.

H. Ensure security training is mandatory for all employees.

2. **Detection**:

A. Determine which devices are affected by the attack and take care of them first. When a ransomware attack is complete, you will usually see a message on the device's screen. If this is the case, take a picture and disconnect the device from the network as soon as possible, preferably immediately, physically, or logically. Do not attempt to turn off the device unless you absolutely must, as this can damage forensic evidence.

B. If you are certain or have a strong reason to suspect that a device may be infected with ransomware, but there is no message yet, physically or logically disconnect it from the network as quickly as possible, ideally immediately. Do not try to turn it off unless you have to, as this can corrupt forensic evidence.

C. If many devices have become infected, it may practically be easier to isolate them all by turning off the network switch or Wi-Fi access point, instead of dealing with each device.

D. Disconnect every network share being used by either confirmed or suspicious devices until the ransomware becomes contained. Try doing this as soon as possible, preferably immediately. Also, include any mirror or disaster recovery versions.

E. Identify what types of data exist on the devices, file shares, or any other systems to which there is a direct connection.

F. Based on this information and the number of devices affected, the severity of the incident is determined. The incident response plan will aid in this identification. If the severity level is uncertain, select a higher severity level, as it can be lowered upon further inspection, but may not receive the attention it needs if it is set too low to begin with.

3. **Triage**:

A. Mobilize the incident response blue team to begin the initial investigation of the cyber incidents. The names of the people on the incident response team should be included in the incident response plan.

B. The easiest recovery solution for ransomware is to restore from a backup. Determine what backups of the affected data are available and validate that these backups are applicable.

C. When external resources are required or public attention is drawn, resources should be mobilized as quickly as possible to seek information.

D. Gather initial data about the incident. At a minimum, it should specify how the incident reported was, what triggered the cyber incident (for example, a lost laptop, suspected hacker, malware, and so on), the location of the data, both physically and logically, the amount of data (for example, the number of accounts, unique numbers, customer names, and so on), whether financial data is included (for example, credit card numbers, pins, expiration dates, and so on), whether personal data is included (for example, names, addresses, ZIP codes, email addresses, and so on), the format of the data (for example, redacted, encrypted, layout, length, and so on), whether the data was encrypted, and if so, how this was provided, a preliminary business impact assessment, and any current action being undertaken.

E.   Back up all artifacts, including copies of data, through secure downloads and screenshots.

4. **Analyze**:

A.   Engage technical staff from the Blue Team.

B.   Determine and investigate whether **Personally Identifiable Information (PII)** (internal or external) or other sensitive data is at risk (if so, use the data loss/theft playbook referenced in this chapter), public or personal safety is affected, services are affected and what these are, critical systems can be controlled/recorded and measured, there is evidence about who is responsible for the attack, and forensic data can be collected (if necessary).

C.   Generate a hash from the suspected file and send it to a sandbox, anti-malware vendor for identification.

D.   Look up other gathered indicators to the identify ransomware variant.

E.   If prior known variant identification is not possible, perform behavioral analysis and reverse engineering. Skip this if identification was possible.

F.   Check Google or other search engines for whether any of the indicators noted previously have been found by other security professionals and note any indicators that they may have listed.

G.   Download any reports by other researchers (if available) for the specific ransomware variant(s) encountered in the incident. Harvest any additional indicators from the report(s). (This report will also be used in the next phase to answer questions.)

5. **Remediation – Contain/Eradicate and Recover**:

A.   If you have not already done so, physically or logically disconnect all suspicious or infected devices from the network.

B.   The ideal solution is to leave the devices turned on but unplugged. In some cases, the ransomware unlock keys remain in memory and can be used to easily recover the device.

C.   Forensic images of the compromised devices may be required to understand the cause of the attack. Optimally, these images are captured *before* mitigation measures are implemented on the system(s); however, this may not always be possible, so you should strive to collect the images in as unaltered state as practicable.

D.   Create a snapshot for virtual systems and ensure that this snapshot cannot be deleted by accident. If possible, add the storage status as well as the dormant data.

E. With physical systems, cloning the physical drives is usually required. Ideally, this is captured while the system is running using a forensics tool; however, an offline clone is acceptable if it is not possible to use a forensics tool. When possible, the original hard drive should be retained and the device recovered with a new hard drive.

F. Gather and examine evidence from additional sources. This may include remotely collected system logs, network device logs (firewalls, IDSs, and so on), blacklist ransomware, and hashes of dropper/infection and persistence mechanisms.

G. Scan endpoints and review system/network artifacts (logs, data streams, and so on) for signs of compromise and lateral movement.

H. Obstruct known communication and infection channels if possible (NAC, email sender, firewall, and so on).

I. Isolate and capture infected hosts (disconnect wireless and wired).

J. If warranted by the scope and severity of the incident, engage an external forensic company to conduct a more thorough review of the impacted systems. This is usually done at the direction of the CSIRT and using one of the pre-identified services that was previously assessed and contracted to provide these services to the organization at large. It is well known that ransomware attacks can be repeated. Therefore, it is important to identify the cause of the infection to reduce the likelihood of such an occurrence.

K. Create a timeline from the incident with as much detail as you can, including answers to the following points: who accessed the systems, and what accounts were modified, accessed, or created? What alterations were the attackers making to the systems? What code, malicious or otherwise, was installed or used by the attackers? What information was compromised, accessed, or exfiltrated? When did the attack take place and for how long did the attackers access the systems? When was the method used by the attackers revealed (if at all)? Where were the affected systems located and what other systems were in the common environment? From where did the attackers access the systems and from where did they penetrate the network and the systems? Why did the systems come under attack; did it involve a targeted attack or just a random attack on a vulnerable system? Why was the attack not detected or stopped by antivirus or other detection tools? How did the attackers gain access to the system; did they take advantage of a misconfiguration, a known vulnerability, or a zero-day vulnerability, or were they intentional actions by a malicious insider?

L. Once the full incident timeline and details that occurred can be determined, the impacted systems can be restored and remediated to get them ready to be put back into service.

M. Compromised systems should ideally never be put back into operation, as there is always the possibility that the attackers could be left behind and threaten the systems. It is best to create replacement systems from scratch and patch all software on these.

N. If the replacement systems are entirely new, it is not necessary to wait for the review to be completed before beginning this part of the process.

O. The only exception is if the exact time of the compromise is known, and a backup copy known to have been clean and created before that time can be used. In that case, the system still needs to be completely patched to remove the attack method before being put back into use.

P. All systems should be hardened per an industry standard to reduce the attack surface and decrease the likelihood of weak or off-the-shelf configurations entering production.

Q. Before deployment, systems need to be scanned for vulnerabilities. If possible, this should be an authenticated scan, as this offers a higher level of security than a simple remote scan. For workstations, a single device can be scanned, provided that all devices are built according to the same standards.

R. When systems are deployed with completely new applications, use the standard organization risk review process for those applications.

S. If a defect (or lack of backups) caused data to be unrecoverable, make sure that your backup process is improved to ensure resistance to ransomware.

T. If a defect in endpoint protection (or lack thereof) enabled or extended the attack, it is important to make sure these vulnerabilities have been addressed and/or mitigated before systems are re-deployed. Optimally, advanced endpoint protection is deployed that utilizes machine learning algorithms and process monitoring, not just signatures, to detect malware.

6. **Post Incident:**

A. Prepare a post-incident report that includes, at a minimum, the following details: details of the cause, impact, and actions taken to help mitigate the cyber incident, including the time, type, and location of the incident, as well as the impact on users; actions taken by responsible blue team members, service providers, and organizational stakeholders to enable the resumption of normal business operations; and recommendations on how any aspect of people, processes, or technology across the organization can be enhanced to help avoid the recurrence of similar cyber incidents as part of a formalized procedure for learned lessons.

B. Establish a formal process for evaluating lessons learned for future preparedness activities.

C. Perform root cause analysis and fix underlying issues.

D. Where appropriate, disseminate the findings to external stakeholders.

E. Communication activities include, among others, publishing internal communications to inform and raise workforce awareness of data breach attacks and apply security awareness and publishing external communications, as appropriate, in line with the communications strategy to advise customers, communicate with the market, and inform the press regarding cyber incidents.

Data loss/theft attacks playbook

In this playbook, it will be worth considering data breaches and one of their impacts, which is data loss/theft. As we saw earlier in this book (see *Chapter 5*, *Threats*), a data breach is an incident, breach of security, or wider privacy violation that leads to the accidental or unlawful destruction, unauthorized retention, misuse, loss, alteration, unauthorized disclosure, or access to data, which was transmitted, stored, or otherwise processed by the organization, its employees, contractors, or service providers.

For this threat, we can have the following sequence of events:

1. **Preparation**:

    A. Review and practice cyber incident response procedures and processes, including both technical and business roles and responsibilities, escalating as necessary to manage major incidents.

    B. Conduct a review of recent cyber incidents and their outcomes.

    C. Review threat intelligence data for threats to the organization, brands, and industry, as well as overall patterns and emerging risks and vulnerabilities.

    D. Ensure access to any necessary documentation is available across the incident response team.

    E. Identify and obtain any needed forensic tools and appliances.

    F. Define threat and risk indicators within the organization's SIEM solution.

    G. Conduct regular awareness campaigns to highlight cyber security risks faced by the employees. It is important to keep all teams vigilant to threats and make sure they report all incidents to the Blue Team.

    H. Ensure security training is mandatory for all employees.

2. **Detection**:

    A. Monitor automated and manual detecting channels, customer and employee channels, and social media for signs of a data breach or compromise – for example, posting customer, employee, or confidential data online; unauthorized third parties contacting customers or their customers with access to personally identifiable or confidential information; targeted emails to customers or employees containing personally identifiable or confidential information; data leakage prevention logs and alerts; missing or theft of devices containing confidential information; lost or stolen paperwork or printouts of data; and other incidents that indicate data has been stolen outside of the network (for example, ransomware).

    B. Report cyber incidents to the incident response Blue Team.

    C. Consider the intelligence value to other partners or organizations affiliated with the organization and share the data collected up to this point.

    D. Where appropriate, report any incidents to the proper authorities (the police, a private security firm, and so on).

3. **Triage**:

   A. Mobilize the incident response blue team to begin the initial investigation of the cyber incidents. The names of the people on the incident response team should be included in the incident response plan.

   B. Determine the likelihood of employee participation and notify HR (for example, of an insider threat).

   C. Collect initial data about the incident that includes, at a minimum, how the incident became reported, what caused the incident (for example, suspected hacker, lost laptop, malware, and so on), the location of the data, both physically and logically, the amount of data (for example, number of accounts, customer names, unique numbers, and so on), whether financial data is involved (for example, credit card numbers, expiration dates, pins, and so on), whether any PII is involved (for example, names, addresses, email addresses, postal codes, and so on), the format of the data (for example, encrypted, redacted, layout, and so on), and whether the data has been encrypted, and if so, how, a preliminary assessment of the impact on the business and any ongoing actions.

   D. Back up all the artifacts, including data copies, by securely downloading them and taking screenshots.

   E. Audit critical systems and identify indications that similar records of data are at risk.

   F. Identify the data's possible sources or owners.

   G. Conduct an initial review of the affected data to assess whether any PII has been compromised.

   H. Research threat intelligence sources and get in touch with the right people in your network who might have faced incidents like this one.

   I. Review the cyber incident's categorization to validate the cyber incident's type as a data loss/theft incident and assess the incident priority based on the initial investigation.

   J. Report data breaches according to the GDPR or other privacy regulations to the relevant data protection officer.

   K. Consider whether reporting suspected or confirmed unauthorized access to any personal data to the authorities is appropriate at this stage.

   L. Create a report that can be shared with your network and make sure that you inform any partners or other businesses the organization is affiliated with.

   M. Consider conducting a full forensic investigation of the incident.

4. **Analyze**:

   A. Certify that all data involved legitimately originated from the organization, is associated with the organization, and is related to the organization, its customers, or its customers' customers.

   B. Perform a thorough technical cyber incident investigation, including, among other things, analyzing network traffic, reviewing security and access logs, as well as vulnerability scans and automated tool results, analyzing suspicious activity, files, or identified malware samples, reviewing anti-malware logs and/or events, without compromising future forensic activities, correlating recent security events or compromise indicators with suspicious activities on the network, identifying the sources of data compromise, and identifying the specific dataset that was compromised and the method of compromise.

C. Determine the attack method and timeline for the cyber incident.

D. Analyze data types and amounts to determine whether a privacy breach has occurred (for example, for PII).

E. Analyze data types and amounts to determine whether a financial data breach has occurred (for example, company financial reports, bank records, customer or employee credit card information, and so on).

F. Analyze data types and amounts to determine whether it exists only in the organization's environments or is shared with (outside) third-party systems.

G. Review the type and amount of compromised data for possible compliance violations.

H. Detect, identify, and report potentially compromised data.

I. Involve data owners and high-level stakeholders to understand the impact of compromised data on the business.

J. Determine the probability that confidentiality, integrity, or availability could have been compromised.

K. Notify high-level business stakeholders of all suspected and confirmed data breaches, which include any unauthorized access to PII.

L. Notify the appropriate authorities, if necessary, by describing the character of the personal data breach, including, to the best extent possible, the classes and estimated number of data subjects and datasets concerned; communicating the name and contact details of the main contact point from which further information may be requested; and describing the probable consequences of this PII data breach.

M. Develop a remediation plan to remediate the problems and damage caused.

N. Involve technical and business impact assessments to build a priority-based remediation plan.

O. Implement a communication strategy and tactic in line with the remediation plan and activities.

5. **Remediation – Containment/Eradication and Recovery:**

A. Contain the data breach technical mechanisms by isolating all affected systems and accounts from the infrastructure by removing them from the network or applying strict access controls to prevent further data exposure; establish rules to prevent detected suspicious traffic from escaping the network; and back up copies of infected systems as well as malware for further examination if this has not already been done in the previous steps. Reverse-engineer malware to detect any indicators of a possible compromise that will help in the eradication phases, secure critical assets and prevent further damage or theft of data, remotely wipe any lost or stolen assets where this is possible, reset passwords of authorized user accounts, restrict privileges where this is possible, isolate unauthorized user accounts, and analyze all stored data.

B.   Limit the business impact of cyber incidents by putting a notification strategy in place, including internal and external notifications; notify employees, service providers, third parties, and customers; help to develop external communications by providing easy-language text per remediation procedure; and engage the data protection authority of the country in which the compromise occurred, as appropriate.

C.   Activities include removing malware identified during the analysis stage by using appropriate tools; removing any identified artifacts that were used to enable the security breach, such as scripts, code, and binaries; disabling system and user accounts that were used as platforms to carry out the attack; identifying common removal methods from trusted sources (antivirus vendors); performing an automated or manual malware removal process using suitable tools; restoring affected network systems from a trusted backup, reinstalling standalone systems using a clean operating system backup before upgrading with trusted backups; changing all compromised account credentials; and confirming policy compliance across the organization's inventory.

D.   From an organization's viewpoint, these recovery actions typically include, at a minimum, recovering systems based on **Business Impact Analysis (BIA)** and business criticality assessment, performing anti-malware assessments and advanced malware scans of all systems across the organization's inventory, resetting the credentials of all affected systems and user account information, re-integrating systems that were previously compromised, recovering damaged or destroyed data, restoring interrupted services, establishing monitoring to detect further suspicious activity, and orchestrating the deployment of necessary patches or vulnerability remediation efforts. All systems that are deployed must be fully patched before being deployed again. It is not good enough to simply patch the software that was the cause of the compromise; all software on the affected system is supposed to be patched. If software cannot be patched, compensating controls must be used to protect the software.

6.   **Post-Incident**:

A.   Prepare a post-incident report that should include, at a minimum, the following details: the details of the causes, the impacts, and the measures taken to help mitigate against the cyber incident, including the time, type, and location of the incident, as well as the impact on users; actions taken by the responsible blue team members, business stakeholders, and service providers to enable the resumption of normal business operations; and recommendations on how any aspect of people, processes, or technology across the organization may be enhanced to prevent the recurrence of similar cyber incidents as part of a formalized process for lessons learned.

    B.   Complete a formal process to evaluate lessons learned for future preparedness activities while providing feedback.

    C.   Perform a root cause analysis and resolve the underlying vulnerabilities and issues.

    D.   Where appropriate, disseminate lessons learned to external stakeholders.

    E.   Communications engagement activities should include, but are not limited to, publishing internal communication messages to inform and educate personnel about attacks on data and to raise security awareness, and, where appropriate, publishing external communications by communications strategies to advise customers, interact with the market, and inform the press about the cyber incidents. However, such communications should include the most relevant information about the cyber incidents while not leaving the organization exposed to attacks or inciting further data breaches.

Phishing attacks playbook

In this playbook, it will be worth mentioning how incident responders can provide appropriate and timely activities in response to a phishing incident or attack. As we saw earlier in this book (see *Chapter 5, Threats*), a phishing attack is an attempt to obtain information such as credentials and credit card information by posing as a trusted company via electronic communication. In a spear phishing attack, an attacker uses knowledge about employees and the organization to make the phishing campaign more convincing and realistic:

1. **Preparation**:

    A.   Review and practice cyber incident response procedures and processes, including both technical and business roles and responsibilities, escalating as necessary to manage major incidents.

    B.   Review past and recent cyber incidents and outcomes.

    C.   Review threat intelligence data to identify threats to the company, brands, and industry, as well as common patterns and emerging security risks and vulnerabilities.

    D.   Ensure that the entire incident response team has access to all required documentation and data (incident response plans, data flow diagrams, and network architecture diagrams).

    E.   Define threat and risk indicators in the framework of the company's SIEM solution.

    F.   Conduct recurring security awareness campaigns to raise awareness of the cybersecurity risks to which employees are exposed. It is important to keep all teams vigilant to threats and make sure they report all incidents to the Blue Team and other stakeholders.

    G.   Ensure security training is mandatory for all employees and done regularly.

2. **Detection:**

    A. Monitor automated and manual detecting channels, customer and employee channels, and social media for signs of a data breach or compromise, including emails with links to external and unknown URLs, spoofed emails, emails that are non-returnable or non-deliverable, messages and flags by users of suspicious emails, messages by external users or customers of any suspicious activity, and notifications and messages from third parties, ISPs, or even law enforcement of suspicious activity assumed. Monitor email using Mimecast, Proofpoint, or any similar technology to ensure phishing emails don't reach the end users.

    B. Report the incident using the relevant communication channels either to the service desk of the organization or through a third party.

    C. Consider whether data loss or a data breach has occurred and if so refer to the data loss/theft playbook mentioned earlier in this chapter.

    D. Classify the cyber incident based on the available information about the phishing attack and the nature of the incident.

    E. Review escalation procedures, processes, and tools; escalate as necessary.

    F. Report the incident to the incident response blue team.

    G. Consider intelligence value to other partners or organizations affiliated with the organization and share the data collected up to this point.

    H. Where appropriate, report any incidents to the proper authorities (the police, a private security firm, and so on).

3. **Triage:**

    A. Mobilize the incident response blue team to begin the initial investigation of the cyber incidents. The names of the people on the incident response team should be included in the incident response plan.

    B. Identify spoofed emails.

    C. Collect initial incident data that includes, at a minimum, the type of cyber incident, how it was reported, the number of users exposed to the phishing email, a breakdown of what caused the cyber incident, the location of the discovery(s) of the incident, both physically and logically, the number of targeted assets throughout the organization (initially), if this is distributed across different organizations, additional reports on the impacted assets, which includes anti-malware logs, network monitoring logs, and system event logs, a preliminary BIA, and any actions that are currently being taken.

    D. Maintain secure artifacts, include copies of the alleged malware, and provide forensic copies of the affected systems to be used for future analysis.

    E. Research threat intelligence sources and consider getting in touch with the right people in your network who might have faced incidents like this one.

F.   Conduct a review of the cyber incident to categorize and confirm the nature of the cyber incident as a phishing attack and evaluate the incident's priority based on the initial investigation.

G.   A full forensic investigation of the incident should be considered.

4.   **Analyze**:

A.   Engage technical staff from the Blue Team.

B.   Determine and investigate whether PII (internal or external) or other potentially sensitive information is at risk (if so, use the data loss/theft playbook mentioned in this chapter), whether any public or personal safety is involved, whether any services are impacted and, if so, what they are, whether critical systems can be controlled/recorded and measured, whether there are indications of who is behind the attack, whether any internal knowledge is involved in causing the incident, and whether criminals could exploit the act.

C.   Identify and determine patch methods.

D.   Scan and review affected infrastructure for signs of compromise resulting from phishing analysis to identify additional vulnerable systems.

E.   Save all supporting evidence to aid in attribution or any anticipated legal action.

F.   Review threat intelligence to identify whether the phishing attack is tailored to target specific individuals or higher-value stakeholders.

G.   Make sure that all infected devices and assets have either been recalled and quarantined or are in the process of being so.

H.   Perform the following identification activities: identify the affected data or systems; identify the compromised user or compromised user credentials; identify the affected IT services; identify the business impact of the attack; identify the spread of the attack throughout the organization; and identify which tools were used to detect the attack.

I.   Involve data owners and high-level stakeholders to understand the impact of compromised data on the business.

J.   Notify senior staff and stakeholders of suspected or confirmed data breaches, including any unauthorized access to PII.

K.   Develop a remediation plan, by performing technical and business analysis, based on priorities and implement a communications strategy following this remediation plan.

5.   **Remediation – Containment/Eradication and Recovery**:

A.   Decrease malicious activity by preventing phishing activities, quarantining and removing affected systems from the network, or applying access controls to isolate them from any production networks.

B.   Deny access to all identified **Remote Access Tools (RATs)** to prevent remote communication with command or control servers, exploited applications, and websites.

C.   Identify compromised accounts or at-risk user credentials.

D.   Identify potentially malicious code and malware on any of the systems that can be connected to the fraudulent site from the phishing attack.

E.   Keep business data owners and stakeholders informed of the progress of containment efforts.

F.   To remediate a phishing attack, the incident response team must be capable of identifying methods of removal based on the results of the phishing attack, performing an automated or manual removal process to eliminate the phishing attack with appropriate tools, and performing a recovery of the affected networked systems using a trusted backup. They must also be able to reinstall individual systems using a clean operating system backup before updating them with trusted backups, modify any compromised account data, and confirm compliance with policies across the organization. They must ensure they implement/reconfigure the anti-phishing solution, as well as ensure they activate multi-factor authentication for all enterprise accounts.

G.   From an organization's perspective, recovery activities include, among others, recovering systems based on BIA and the criticality to the business, performing anti-malware and extended malware scans of all systems across the organization's inventory, resetting the login credentials of all affected systems and user account information, reintegrating systems that were previously compromised, recovering corrupted or destroyed data, restoring all interrupted and suspended services, establishing monitoring to identify and detect any further suspicious activity, and orchestrating the deployment of all necessary patches or vulnerability remediation efforts. All systems that are deployed must be fully patched before being deployed again. It is not enough to patch only the software that was the cause of the compromise; it is recommended that all of the system's software is patched. If some software is not patchable, compensating controls must be used to protect the software.

6.   **Post-Incident**:

A.   Prepare a post-incident report that should include, at a minimum, specifics on the details of the cause, impact, and actions taken to help mitigate the cyber incident, including the time, location, and type of incident, as well as the impact on users; actions taken by the responsible blue team members, business stakeholders, and service providers to enable the resumption of normal business operations; and recommendations on how any aspect of personnel, processes, or technology across the organization could be enhanced to help avoid the recurrence of similar cyber incidents as part of a formalized process for lessons learned.

B.   Establish a formal process for evaluating lessons learned for future preparedness activities.

C.   Perform root cause analysis and fix underlying issues and vulnerabilities.

D.   Where appropriate, disseminate the findings to external stakeholders.

E. Communication activities include, among others, publishing internal communications to inform and raise workforce awareness of data breach attacks and apply security awareness, publishing external communications, as appropriate, in line with the communications strategy to advise customers, communicate with the market, and inform the press regarding cyber incidents.

Disaster recovery planning

Disaster recovery plans have almost the same structure as incident response plans, but let's see it on a part-by-part level again.

A disaster recovery plan consists of the following parts:

- Purpose
- System/application name and description
- Recovery parameters
- Recovery team
- Definitions
- Disaster recovery plan procedure
- Revision history

First, as with the incident response plan, there is an introduction stating the purpose of the document. This can give the teams a better understanding of what the plan is and which system or application it refers to. There may be more than one plan in an organization, as with incident response plans.

Second, there is a better explanation of the system, or the application referred to in the disaster recovery plan. There can be an identification name for the application/system that correlates to the asset inventory or other documents that have application references. In this part, there should also be an application or system name and finally a description of the application or system regarding what it normally does in the organization.

In the next part of the plan, there are recovery parameters. Those parameters will refer to other documents that should have been completed by the Blue Team or other teams that work with the system. There should also be a risk profile of the system or application that contains the nature and level of threats faced by an organization, as we saw in *Chapter 3, Risk Assessment*. Next, there should be a BIA, which should give the responders an understanding of the business impact that the organization will face if a disaster occurs.

The next part contains the recovery team members, along with their emails and phone numbers. The first name stated on the document should be the technical lead of the application or system and should be contacted first if a disaster occurs, along with other members of the recovery Blue Team.

The definitions part should contain, as stated in the *Incident response planning* section, all the definitions of different words or combinations of words mentioned in the document thus far.

The sixth part should contain the procedure of the disaster recovery plan. This can be depicted as a diagram with steps that the recovery teams should follow. This can also contain if statements, which should lead to different solutions according to whether those statements are true or false. Following the flow of the diagram should help blue teams decide what the best solution is for disaster recovery. Each of those parts should be written down and explained. Each step in the flow diagram must be completed before disaster recovery occurs. A disaster recovery diagram can be seen in *Figure 10.1*:

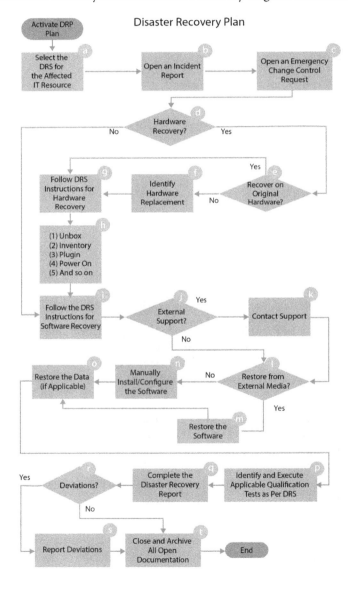

Figure 10.1 – Disaster recovery diagram

Last but not least, same as with incident response plans, there is the version history, which contains the version number of the document, along with the author's name, the date of publication in the internal system, and what revisions were made to the plan.

A good example of this plan is data sabotage. One type of data disaster that can strike an organization is when someone – an angry employee, for example – intentionally commits data sabotage. For example, the employee might insert inaccurate or falsified data into the organization's databases to reduce data quality and make the data useless to the organization. The employee could even insert some malicious code into the company's data to deliver malware to the systems.

However, the crucial step in preparing for this type of event is to ensure that there are backups of the data that extend back far enough to allow recovery with a version of the data that is considered safe. If the only available copy of the data was made a day ago, but the damage took place 3 days prior, the backup copy will not be helpful in this situation. For this reason, it is advisable to create multiple backups of the organization's data, if possible. This way, data from a week ago can be restored in case newer backups contain corrupted information. The disadvantage of using an older backup for disaster recovery, of course, is that any data that has been added or changed since the backup was created will be discarded. However, in certain recovery situations, this is the price an organization will have to endure.

It is also worth considering in such an example that if the organization can determine which parts of its data have been tampered with, the recovery team can leave the necessary data intact and recover only the corrupted data to help minimize data loss.

A further example of a disaster is the destruction of a data center. Among the worst-case scenarios that a modern organization can face is a disaster that destroys a portion or all of its data center, including all servers and hard drives within it. Although a situation like this is rare, it can still occur, and not just as a result of a major natural catastrophe such as a hurricane or earthquake. Problems such as power surges and even squirrels are capable of inflicting permanent damage on data centers. Squirrels damage electrical distribution facilities by tunneling, by chewing through electrical insulation, or by simultaneously coming into contact with two conductors at different electrical potentials.

For an organization, the best way to prepare for recovery from such an event is to ensure that offsite copies of the data are in place. When the organization's production data is onsite, as in the hospital example in *Chapter 3, Risk Assessment*, and this catastrophe occurs, the hospital should have made sure the backups are stored in a different data center or the cloud. If the data is hosted in the cloud, then the team could back it up to local storage or another region hosted in the same cloud. The organization should also ensure there is a way to quickly restore the backup data to a new infrastructure. Transferring high volumes of data from one location to another using the internet can take considerable time, so it's not always a smart strategy in the disaster recovery context. Sometimes, it may be more rapid to physically move backup copies of hard disks from servers located in the data center where the backup is located, and then to connect and transfer the backup data to the organization's production servers and infrastructure.

Recovering data from a data center that's been destroyed requires offsite copies of the data, as well as a plan to quickly get that data to where it is needed to keep the business running after a disaster.

As another example, a lengthy network disruption can be just as disruptive as a loss of data. If the network breaks – or if even a single workstation client can suddenly no longer connect – IT managers need to react swiftly.

Network disruption preparedness testing is the best way to verify that the organization can quickly resolve problems when they arise. Several network testing tools are available to help with simulating commonly encountered disaster scenarios.

Example tests include the following:

- Testing for unexpected surges and spikes in network traffic
- Mock tests replicating the effects that a crippling network attack has
- Network health testing to identify potential problems in specific areas of the network
- Readiness tests to ensure that IT teams are capable of a rapid response

These tests should not be confined to software-based testing only. Network administrators must be routinely testing and going through the recovery protocols for these disaster recovery scenarios to ensure they know precisely what needs to be done in the event of a real-world disruption.

Another example would be a series of disaster scenarios that can be extremely damaging to your employees and your operations – and can still have little to do with your IT systems. For this reason, disaster recovery testing (and business continuity testing) must not be limited strictly to IT.

What if the organization is targeted in a workplace shooting? How would employees need to protect themselves? Where can they seek safety?

By testing various crisis scenarios, the risk of damage to the most valuable asset – people – can be significantly reduced. And by protecting employees, the organization is also protecting its operations.

There are a few tests to consider regarding this type of disaster:

- Evacuation drills for fires, and other on-site dangers, such as active shooters
- Emergency procedures for earthquakes, tornados, and other natural disasters
- Testing the communications systems that will be used to keep employees updated during a prolonged disaster

A fire drill is perhaps the most common form of testing for an on-site emergency and in some areas, these drills are required by law for certain types of commercial buildings. However, fires aren't the only scenario that employees should be prepared for, especially in larger buildings.

Routine training should be conducted to educate employees on how to safely respond to all the on-site dangers identified in your disaster recovery plan. When employees know what to do in an emergency, they are at far less risk of harm. That's good for their well-being, of course, but it's also good for the business.

In the next section, we will focus on cyber insurance and how this can help a business stay afloat after an incident or disaster.

Cyber insurance

Cyber security insurance, also referred to as cyber insurance or cyber liability insurance, is a deal that an organization can take out to mitigate the financial risks involved in online business. For a monthly or quarterly fee, the insurance transfers a portion of the risk to the insurance provider.

Cyber insurance is a new and emerging industry. Companies that purchase this type of cybersecurity insurance are now considered to be early adopters. Cyber security policies can change month to month as the associated cyber risks are dynamic and fluctuating. Unlike established insurance plans, cybersecurity insurance policy insurers only have limited data to shape risk models to establish insurance coverages, rates, and premiums.

Cyber insurance has its genesis in **Errors and Omissions (E&O)** insurance, a category of separate insurance that protects against errors and defects in the services provided by an organization. E&O insurance is similar to product liability insurance for companies that are selling products, either physical or digital.

Some cyber insurance policies do include specific provisions for E&O coverage, but most providers sell these as separate and standalone policies. E&O insurance will not cover third-party data loss, such as customer bank account information; clients who need such protection can opt for a cyber insurance policy that will cover it.

To get cyber-insured, an organization must get a policy from an insurance agent and make sure their insurance covers the following:

- Data breaches (such as incidents involving theft of personal information). This is because without it, if an organization loses data, then there is no coverage for that loss.

- Cyber attacks on the data held by third parties.

- Cyber attacks (such as breaches to the organization's network).

- Cyber attacks that occur anywhere in the world (not just in the business's country of origin).

- Terrorist acts.

It is also worth considering that the cyber insurance provider will do the following:

- Defend the organization in a lawsuit or regulatory investigation (**duty to defend** wording)

- Provide coverage in excess of any other applicable insurance that the organization has in place for other situations

- Offer a breach hotline that's available every day of the year at all times

In cyber insurance, there can also be two types of coverage: first-party cyber coverage and third-party cyber coverage.

First-party cyber coverage protects the organization's data, including employee and customer information. This coverage should typically include the business costs related to the following:

- Legal counsel to determine the organization's notification and regulatory obligations.
- Recovery and replacement of lost or stolen data.
- Customer notification and call center services.
- Lost income due to business interruption (if a cyber attack incurs a loss of business, the organization taking part in the insurance policy will have their losses covered).
- Crisis management and public relations. Crisis management is not like disaster recovery, even though they can be connected. If, for example, the organization works with people and the people cannot access the application or the website, then the cyber insurance agent will contact them to alleviate the crisis, which might be caused by the application malfunctioning.
- Forensic services to investigate a breach. Forensics was covered in *Chapter 9, Cyber Threat Intelligence*. This clause could mean that the agent will have to perform forensics on the situation that an organization is facing.
- Charges, penalties, and fines related to the cyber incident.

Third-party cyber insurance coverage usually guards an organization against liability if a third party makes a claim against it. This coverage commonly includes the following:

- Payments to affected consumers of a data breach
- Claims and expenses for the settlement of disputes or lawsuits
- Losses associated with defamation and infringement of copyrights or trademarks
- Litigation costs and costs of responding to inquiries from regulatory authorities
- Other claims settlements, compensation for damages, and court judgments
- Accounting expenses

As mentioned at the beginning of this section, cyber insurance has its flaws. In 2020, the world entered a new era of cyber attacks. Although before that there were many decades of malware, breaches, and other forms of attacks, during 2020, the world saw increased bad actor sophistication, prosperity to pay in ransomware cases, and a broad swath of geopolitical uncertainty – conditions that hackers found advantageous to their plans.

The severity of the financial consequences was profound. Ransoms rocketed from five-figure price tags into the millions. A report by Hiscox indicated insured cyber losses of 1.8 million dollars in 2019, up an eye-popping 50% year over year.

Facing the prospect of financial fallout from an attack, organizations around the world turned to cyber insurance. In 2020, the global insurance community saw the first cyber insurance program to exceed 1 billion dollars!

However, the momentum that propelled this sector of cybersecurity thus far may be running out. This is because businesses look at cyber insurance as a luxury and not a necessity, as the economic strain of the Covid-19 pandemic has shown. While more attacks could stimulate demand, they also create a problem in the supply of that insurance, making insurance agents warier of providing cover and reinsurers (who provide insurance for insurance providers) less interested in backing cyber liabilities. Not only that but also the lack of historical loss data adds another layer of unpredictability for all involved.

Determining how much cyber insurance is needed should be a risk management Blue Team job. As blue teams conduct risk assessments, they can determine the need for cyber insurance; whatever they can't cover in the cybersecurity landscape with their controls, they can cover with cyber insurance.

But as mentioned many times in this book, attackers are unpredictable, and we never know whether the controls that were placed during the risk assessment will fail their prime function when they are needed most. Zero-day attacks are also a prime reason why cyber insurance should be a necessity and not a luxury as it is considered today.

For example, in 2011, Sony's PlayStation Network was breached by hackers, exposing the PII of 77 million PlayStation user accounts. The breach prevented users of PlayStation consoles from accessing the service, an outage that lasted for 23 days. Sony incurred over $171 million in costs related to the breach. Portions of this cost could have been covered by a cyber insurance policy, but Sony did not have one in place. A court case ruled that Sony's insurance policy covered damage to physical property only, leaving Sony to incur the full amount of costs related to cyber damages.

It has been noted that every threat that is out there can be controlled by enlisting the right measures, those being controls and other types of procedures.

Summary

In this chapter, we understood incident response and disaster recovery measures, how they are conducted, and examples of incidents and disasters covered and not covered by a plan. We also noted that a cyber insurance policy is a must for organizations nowadays, and we provided examples of where cyber insurance has failed or been a necessity. In the final chapter, we will conclude this book by learning how to prioritize and implement a blue team strategy in an organization.

11

Prioritizing and Implementing a Blue Team Strategy

Throughout this book, we've seen all the aspects of blue teams, how they are organized, how they conduct their everyday activities, and how they plan and make sure they have covered every eventuality – we've analyzed how they do so exclusively.

In this chapter, we will analyze emerging technologies in the detection and prevention space for blue teams, and we will also analyze any emerging strategies for blue teams and much more.

This chapter will summarize the different concepts covered in this book via the story of a blue team's creation, and you will come to understand the need to prioritize a blue team strategy in your organization.

In this chapter, we will cover the following:

- Emerging detection and prevention technologies and techniques
- Pitfalls to avoid while setting up a blue team
- Getting started on your blue team journey

Emerging detection and prevention technologies and techniques

In this section, we need to understand that prioritizing and implementing a blue team requires us to look into the different ways that this blue team can be utilized. We will be mentioning some emerging detection and prevention technologies and techniques, including the following:

- Adversary emulation with some examples
- A **Virtual Chief Information Security Officer** (**VCISO**)
- Context-aware security
- Defensive AI
- Extended detection and response
- Manufacturer usage description
- Zero Trust

Adversary emulation

Adversary emulation is a defensive technique that uses simulated attacks in order to train the blue team.

This can provide blue teams with actionable data that helps them uncover and resolve vulnerabilities and security issues.

It also allows them to use controls and solutions already implemented in their organization, along with their capabilities to detect and prevent malicious attackers and suspicious behavior.

As Sun Tzu states in *The Art of War*, and as applies to this technique of adversary emulation, "*If you know the enemy and know yourself, you need not fear the result of a hundred battles. If you know yourself but not the enemy, for every victory gained you will also suffer a defeat. If you know neither the enemy nor yourself, you will succumb in every battle.*" Knowing your enemy is really important to security – knowing yourself means that you know what you can do when the enemy strikes.

This technique helps incident responders understand the mindset of an attacker and by doing so, they can plan their response. This also helps them learn how to cover their backs. A blue team strategy is about building a landscape in which a decision should be made as quickly as possible. It is not a question of if the organization will be attacked but when it will be attacked.

Have a look at these links for some open source examples of these technologies:

- **APT Simulator**: `https://github.com/NextronSystems/`
  `APTSimulator?rel=nofollow,noopener,noreferrer&target=_blank`

- **DumpsterFire**: `https://github.com/TryCatchHCF/ DumpsterFire?rel=nofollow,noopener,noreferrer&target=_blank`

- **caldera**: `https://github.com/mitre/ caldera?rel=nofollow,noopener,noreferrer&target=_blank`

VCISO services

An emerging new service technique that can help start-ups or emerging businesses without hiring a full-time CISO is having a VCISO.

A VCISO cannot replace a blue team but they can help start a conversation with executives in a company and provide risk assessments for an organization, along with consultation for building an effective cybersecurity and resilience program. This kind of service can also facilitate the integration of security into a business strategy, process, or culture. VCISOs can help with the integration and interpretation of information security program controls. Finally, they can serve as security liaisons to auditors, assessors, or examiners.

Because we are not advertising anyone in particular, VCISOs can be based anywhere around the world – they just need to be considered a part of the teams or organizations they protect.

A VCISO shouldn't be a permanent solution but a temporary one until an actual CISO is hired by an organization. However, it can be considered a good solution for organizations that are starting up a business plan.

VCISOs can be found on LinkedIn or, if they are part of a bigger group, in an enterprise capacity.

Context-aware security

Context-aware security is a type of technology that helps businesses make security decisions in real time.

Traditionally, cybersecurity technologies assess whether or not to allow someone access to data or a system by asking yes/no questions. This simple procedure can cause some legitimate users to be denied, thus slowing productivity.

Denying entry to an authorized user is reduced when implementing context-aware security. Instead of relying on answers to yes/no questions, supportive information such as time, location, and URL reputation to assess whether a user is legitimate or not is used by context-aware security.

Context-aware security streamlines data-accessing procedures and makes it easier for legitimate users to do their work. However, one concern for users of this technology may be end user privacy since they are monitored regularly and that does not allow them to have privacy in their day-to-day activities.

Defensive AI

Defensive AI can be used to detect and stop cyber attacks. Cybercriminals tend to use technologies such as offensive AI and adversarial **Machine Learning** (**ML**) because they are more difficult for traditional cybersecurity tools to detect.

Offensive AI includes deep fakes, false images, personas, and videos that convincingly depict people or things that never happened or do not exist. Adversarial ML can also be used by attackers to trick machines into malfunctioning by giving them incorrect data.

Defensive AI can be used by a blue team to detect and stop offensive AI from measuring, testing, and learning how a system or network functions.

Defensive AI can boost algorithms, making them more difficult to break. Harsher vulnerability testing can be conducted on ML models by the blue team.

Extended Detection and Response (XDR)

XDR is a type of advanced cybersecurity technology that detects and responds to security threats and incidents. XDR can respond across endpoints, the cloud, and networks. It evolved from the traditional endpoint detection and response, as we've seen in *Chapter 8, Detective Controls,* and *Chapter 10, Incident Response and Recovery.*

A more holistic picture is provided by XDR, which makes connections between data in different places. Threats can be detected and analyzed by the blue team from a higher, automated level. Current and future data breaches can be prevented or minimized across an organization's entire ecosystem of assets.

XDR can be used by the blue team to respond to and detect targeted attacks, automatically confirm and correlate alerts, and create comprehensive analytics. Benefits of XDR include the automation of repetitive tasks, strong automated detection, and a reduction of the number of incidents that need investigation.

Manufacturer Usage Description (MUD)

The **Internet Engineering Task Force** (**IETF**) created a standard called **MUD** in order to strengthen the security of IoT devices in small businesses and home networks.

Network-based attacks are a prime vulnerability to IoT devices as we saw in *Chapter 3, Risk Assessment.* IoT devices need to be secure without big costs or being too complicated.

Using MUD includes benefits such as simple, affordable, and improved security for IoT devices. Blue team members can use MUD to make devices more secure against **Distributed Denial-of-Service** (**DDoS**) attacks. The amount of damage and data loss can be reduced by using MUD after a successful attack.

Zero Trust

As we saw in *Chapter 4, Blue Team Operations*, and *Chapter 7, Preventive Controls*, traditional network security follows the motto "*trust but verify*," which assumes the users within an organization's network perimeter are not malicious threats. Zero Trust, on the other hand, aligns itself with the motto "*never trust, always verify*."

In a zero-trust network security landscape, all users authenticate themselves before they can access an organization's data or applications. It is not assumed that users inside the network are more trustworthy than anyone else. This stricter scrutiny of all users can result in greater overall information security for the organization.

Blue teams can use zero trust to deal more safely with remote workers and challenges such as ransomware threats. Various tools may be combined in a zero-trust framework, including multi-factor authentication, data encryption, and endpoint security.

The execution of this framework combines advanced technologies, such as risk-based multi-factor authentication, identity protection, next-generation endpoint security, and powerful cloud workload technology, to verify the identity of a user or system, examine access at that time, and maintain system security. Zero trust also requires considering data encryption, securing emails, and verifying the safety of assets and endpoints before connecting them to applications.

As a result, it must be ensured that the blue team vets all access requests continuously prior to permitting access to any organizational assets. That's why the implementation of Zero Trust policies relies on real-time visibility into many user and application identity attributes such as the following:

- User identities and credentials
- Credential privileges on each device
- Normal connections for the credential and device (behavior patterns)
- Endpoint hardware type and function
- Geo-location
- Firmware versions
- Authentication protocols and risks
- Operating system versions and patch levels
- Applications installed on endpoints
- Security or incident detections, including suspicious activity and attack recognition

Events must be tied to the use of cyber analytics, broad enterprise telemetry, and threat intelligence in order to ensure better algorithmic **AI/ML** model training for a hyper-accurate policy response. Organizations should thoroughly assess their IT infrastructure and potential attack paths to contain attacks and minimize any impact if a breach should occur. Segmentation by device types, identity, or group functions can be included in this. For example, suspicious protocols such as **Remote Desktop Protocol (RDP)** or **Remote Procedure Call (RPC)** to the domain controller should always be challenged and should not be used by the organization.

The use or misuse of credentials in the network occurs in a lot of attacks that the blue team defends against. With constant new attacks against credentials and identity stores, additional protections for credentials and data extend to email security and secure web gateway providers. Therefore, greater password security, the integrity of accounts, adherence to organizational rules, and the avoidance of high-risk shadow IT services must be ensured.

Security professionals have increasingly formalized zero trust, although it has been described as a standard for many years as a response to securing digital transformation and a range of complex, devastating threats that appear in the blue team's work every day.

While any organization can benefit from Zero Trust, an organization can benefit from Zero Trust immediately if the organization is required to protect an infrastructure deployment model that includes the following:

- Multi-cloud, hybrid, or multi-identity

- Unmanaged devices

- Legacy systems

- SaaS apps

The organization needs to address key threat cases, including the following:

- Ransomware – a two-part problem involving code execution and identity compromise

- Supply chain attacks – typically involve unmanaged devices and privileged users working remotely

- Insider threats – it is especially challenging to analyze behavioral analytics for remote users

According to their business, every organization has a unique digital transformation maturity, current information security strategy, and challenges. Zero Trust, if implemented properly, can be adjusted to meet an organization's specific needs, and still ensure a return on investment on your blue team strategy. In the next section, we will be mentioning situations to avoid when setting up your blue team.

Pitfalls to avoid while setting up a blue team

The most common way intelligence teams are formed is this: leadership decides their organization needs a blue team. Those in charge have either come to this conclusion because they were influenced by their peers or more often, those above them informed them that it was a need. Although there's nothing inherently wrong with leadership deciding the organization needs a blue team, this top-down approach is extremely vulnerable to common mistakes that can undermine the credibility and efficacy of a blue team. However, the decision has been made, the die has been cast, and team building begins.

In these cases, the most common action is to promote a high performer from within – who has no background in information security – to lead the new blue team. That person, often carrying a strong incident response or network security background, attempts to fill the role by either falling back on what they know or reading as much as they can to learn about what a blue team is, on the fly.

Frankly, information security is not a hobby. It's not a subset of cybersecurity to which one can easily pivot. Nevertheless, they try, and the team they build usually looks a lot like themselves. The enterprise ends up with a team of **Security Operations Control** (**SOC**) analysts and incident responders with titles unrelated to information security. It's not that the team isn't skilled. They just happen to lack the particular skill sets needed for this very specific function. This is just one example of good people trying to do the right things while being in positions in which they are unlikely to succeed.

Ultimately, when this type of team is unsuccessful at making the organization any more proactive and misses the mark on measurable goals and objectives, everyone updates their resume and looks for new jobs in their old specialties. Or, worse yet, these people now market themselves as blue team analysts because there is a massive gap in the market, and they now have that title on their resumes.

Despite their best efforts to address the information security requirements of a customer that does not know how to address them, vendors often take the rest of the blame for the failure to build an effective blue team program. Therefore, a whole new batch of vendors, happy to capitalize on the perception of their competition failing, will be the benefactors of this directional change – but if this talent strategy doesn't change, the results are unlikely to improve.

Another common way agencies create blue teams is by hiring personnel from the **Intelligence Community** (**IC**) and law enforcement agencies. The rationale is that there is a lot of impressive talent in government and that these people bring experience and credibility. Who wouldn't be impressed by a team filled with hundreds of years of experience in three-letter agencies, right? It's true. However, there are two serious challenges to this approach:

- **Culture shock**: People who spend entire careers inside the government can become institutionalized. They can struggle to familiarize themselves with an entirely different set of goals, expectations, budget plans, schedules, and social standards.

- **Verification**: Far too many people are coming out of the IC with imposing resumes that are hard to validate. They hide behind "*it's classified*" knowing most will not check. *NEVER* hire someone unwilling or incapable of verifying their credentials.

A team constructed entirely on amazing credentials in intelligence or law enforcement will also likely struggle to create enough variety of thought, flexibility, and adaptability. This kind of *groupthink* can result in organizational confirmation bias – an echo chamber – that leads to inaccurate conclusions.

No matter how much we invest in access methods, different tools, or cutting-edge technologies, information security is still about what people we have. Therefore, as you build your blue team program – and evaluate all those vendors – don't overlook the importance of investing in the right people and partners. Otherwise, you can find yourselves on a treadmill of changing personnel and vendors that give the appearance of progress while getting no closer to the stated blue team objective.

In the next section, we will discuss how to start a blue team in your organization by summarizing some points from the previous chapters of this book.

Getting started on your blue team journey

As discussed in this chapter and *Chapter 2, Managing a Defense Security Team*, the first and most important thing that an organization can do is hire a CISO or employ a VCISO's expertise in order to kickstart a blue team's strategy.

As we learned in *Chapter 3, Risk Assessment*, creating an inventory of assets is the first thing a CISO should do. The CISO will perform the first risk assessment next in order to try to calculate the risk that the organization faces in the near future.

Subsequently, they will start hiring for the risk management blue team, getting enough people to help implement a **Risk Management Framework (RMF)**. According to *Chapter 4, Blue Team Operations*, the next thing would be to make the organization security-aware. The risk management team will conduct sessions with the business team and teach them about security awareness until a new staff for that exact purpose has been hired.

As mentioned in *Chapter 6, Governance, Compliance, Regulations, and Best Practices*, the team should not forget about regulations, compliance, assurance, and risk management in order to govern its procedures securely and facilitate the management of the compliance blue team, which should already be present within the organization.

As discussed in *Chapter 7, Preventive Controls*, and *Chapter 8, Detective Controls*, they would then place controls where needed, thus protecting the infrastructure and the people in the organization.

Afterward, as we've seen in *Chapter 9, Cyber Threat Intelligence*, and *Chapter 5, Threats*, they would be able to use threat intelligence to delve deeper into what threats they may face in the coming year. Then, likely discovering that their team will not be enough to face the storm of attacks that may come, they would hire a SOC team.

As deliberated in *Chapter 10, Incident Response and Recovery*, the blue team needs to be trained in their incident response methodology so that they know how to be proactive regarding incidents and disasters by using one of the techniques mentioned in this chapter – adversary emulation or other tabletop exercises that explain how attacks happen.

Summary

For some people, technology has become magic – they know it works but have no idea how. Those who control this magic fall into two categories, protectors (blue teams), exploiters (red teams and unethical hackers), and those that do both (purple teams). Society uses technology to store and transfer more and more valuable information every day. It has become the core of our daily communications, and no business can run without it. This dependency, as well as technology's inherent complexity, has created ample opportunity for those that deem harm to an organization to exploit technology to their advantage. It is each organization's responsibility to ensure that the blue team, its protectors, not only understand how they can protect themselves but also how to successfully respond to, investigate, and help prosecute attackers as they appear.

We sincerely hope that this work has opened your eyes to all the fearsome threats that are out there. We must understand the blue team code, in which we protect what is owned by our organization by being aware and making sure that we have covered our backs with the controls or the people working in the team, who are always vigilant and want to keep us safe within a dangerous cyber world. Keep your eyes open to the horizon – this world of IT has many things to offer, but never forget that as with all good things, bad things can happen. We hope that this book has helped you understand the need for a blue team strategy in your life and your organization. It all starts with the little things and then moves on to the big things. Keep safe, feel protected, and fight the evil that is out there.

Remember, there is no single strategy that can fit every organization, there is no silver bullet, and hence this book should be considered a guide, not a solution to solve the everlasting information security problem.

Part 3:
Ask the Experts

They say setting up a good blue team is as much an art as it is a science. Hence, in this part of the book, we change gears and talk to industry veterans to understand how they have gone about building cyber defenses, what challenges they faced, and, most importantly, what we can learn from them.

In this section, you will read about the experiences of the following people:

- *Anthony Desvernois*
- *William B. Nelson*
- *Laurent Gerardin*
- *Peter Sheppard*
- *Pieter Danhieux*

Each of these experts was given the same prompt: *Please share your best practices and/or the challenges you faced on your journey of setting up a cyber blue team. What advice could you share with our readers?*

The spirit of the prompt is to help you understand the pitfalls to avoid when you embark on your own journey to build a cyber blue team.

On the same note, it is also important to understand that no two organizations are the same. What worked for one organization may not necessarily work for another. Every blue team must understand its organization, the core business, and also the culture of the organization, well before embarking on strategizing and building the required controls for its environment.

Always remember to run a detailed risk assessment in your own organization to understand the needs, the appropriate controls, and the right areas to focus cyber defenses on.

12
Expert Insights

Anthony Desvernois

As an IT security manager, I led security departments in financial institutions in Europe, the Asia-Pacific region, and the Americas. I managed business security teams and operational security teams, covered business and IT continuity topics, and participated as a blue team member in multiple engagements (including purple team exercises).

The blue team is an important part of your cybersecurity team. It is in charge of detection and often the incident response too. They need to maintain constant vigilance in order to be able to anticipate and discover attacks on your information system.

The biggest challenge is to hire *good* people for your team. As Aristotle said, "*The whole is greater than the sum of its parts*," and that totally applies when you need to build a strong team, especially in the case of a blue team. IT security is a wide topic – you can't cover it just with some highly specialized individuals in the long run. You want a lot of energy, esprit de corps, and synergy within your blue team. Therefore, aiming for the most skilled resources is not necessarily the best approach. A better approach is to choose the ones with a strong team mindset and willingness to learn and share. Blue team success is measured over time, so having the right people to go the distance is key.

One of my most skilled blue team members didn't have any past experience with the topic; he just showed genuine interest and had proven teamwork experience and the capability to learn. On the contrary, I once hired a *rockstar* on the team; even if he was very nice and skilled, he didn't stay for long because he lacked the opportunity to express his personality and do his best as an individual contributor. When he left, almost all of his expertise went with him because he couldn't pass those skills on to others.

In terms of best practices, my advice would be to build a strong relationship with the red teams or the ones driving them. That's the way to get real value for the entire organization. At the end of the day, what matters is not to *win* the exercise but to learn the most from it and improve your security posture.

On that topic, we always tried to create a friendly environment and organized lunches and drinks after and during the engagement in order for all participants to share their tips and tricks, share their experiences from past engagements, and also learn about how other companies are organizing themselves. There is a lot to be gained, and it also helps to expand your professional network of security practitioners.

In conclusion, the blue team is a lot about people, and you really need to invest in them if you want to be successful.

William B. Nelson

Bill Nelson is the founder and chair of the **Global Resilience Federation** *(GRF)* (https://www.grf.org/), *a multi-sector non-profit association dedicated to helping ensure the resilience and continuity of organizations against cyber and physical threats, incidents, and vulnerabilities. The GRF is headquartered in Herndon, Virginia.*

Career

At the GRF, Nelson has led the growth of information-sharing communities that make up a network that shares threat, vulnerability, and incident data within their respective sectors and on a cross-sector basis that results in the detection, response to, and mitigation of those threats. The GRF serves many industry sectors, including energy, manufacturing, utilities, operational technology, aviation, auto, space, financial services, K12 school districts, law firms, retailers, healthcare, and accounting and consulting firms.

Nelson also formed and now serves on the board of the **Operational Technology Information Sharing and Analysis Center (OT-ISAC)** (https://www.otisac.org/).

The OT-ISAC is a Singapore-based non-profit company with the mission to protect global operational technology assets in multiple sectors, including maritime, energy, shipping, manufacturing, and others.

He also helped found the **Business Resilience Council (BRC)** (https://www.grf.org/brc), a non-profit company created to foster sharing and cooperation regarding significant incidents, threats, and businesses that impact operations. The **Operational Resilience Framework (ORF)** Work Group of the BRC has developed a framework for the immutable recovery of data, systems, networks, devices, applications, and configurations in response to destructive events.

Before joining the GRF in 2019 as chair and CEO, Nelson was the president and CEO of the **Financial Services Information Sharing and Analysis Center (FS-ISAC)** (https://www.fsisac.com/) from 2006 to 2018. During his 12 years at the FS-ISAC, he grew the membership from under 200 companies to over 7,000 organizations in 50 countries. As the head of the FS-ISAC, he helped coordinate the financial services industry response to major cyber-attacks, including account takeovers, **Distributed Denial-of-Service (DDoS)** attacks, business email being compromised, ransomware, and destructive malware.

Under his leadership, the FS-ISAC collaborated with Microsoft to remove the Zeus botnet infrastructure (`https://news.microsoft.com/videos/operation-b71-microsoft-and-financial-industry-battle-the-zeus-botnets/`).

Nelson has served on multiple boards of directors and is a frequent speaker on risk management (`https://www.cnbc.com/2014/12/18/could-a-cyberattack-take-down-a-bank.html`), cybersecurity, information sharing (`https://www.youtube.com/watch?v=Q8ZiAoaxw2U`), payments, and business resilience (`https://issuu.com/stratton/docs/tt_september`). He has also testified on cyber threats (`https://www.c-span.org/video/?446065-1/banking-hearing-focuses-financial-cybersecurity-risks-preparedness`) and information sharing to both houses of the U.S. Congress. He has received several awards, including the RSA Award for Excellence in Information Security (`https://www.rsaconference.com/experts/bill-nelson`), the NACHA Payments System Excellence Award, and the FS-ISAC Critical Infrastructure Protection Award.

Non-profit and volunteer work

Nelson currently serves as a volunteer at a number of nonprofit companies, including as director of the Operational Technology ISAC, the chair of the GRF, and a member of the Leadership Council of the **National Small Business Association (NSBA)**. He has previously been on the board of Sheltered Harbor (`https://shelteredharbor.org/about`), a non-profit subsidiary of the FS-ISAC with the purpose of protecting customers, financial institutions, and public confidence in the financial system in the case of a catastrophic event.

As CEO of the FS-ISAC and then at the GRF, I learned about the importance of information-sharing communities as a key component in defending against cyber-attacks. I also learned that exercises were especially effective in preparing for response and recovery from future real-life attacks.

The first major exercise we conducted at the FS-ISAC was the **Cyber Attack Against Payment Systems (CAPS)** exercise that was started in 2010. CAPS started out as a 3-day virtual exercise and over the years, transitioned into a 2-day exercise with daily injects of new attack vectors added to simulate a dynamic attack environment. The most challenging aspect of this exercise was the development of attack scenarios that participants would find relevant to their particular environments and that they could take away lessons from to improve their company's cybersecurity. By making the CAPS exercises virtual, we grew to have over 2,000 financial institutions from around the world participate.

Another major initiative that I led was the development of a cyber range, which was a hands-on keyboard-type of blue/red team exercise. The cyber range allowed participants to exercise their defense tactics in a real-time simulated environment against various types of malware and DDOS attacks. We worked closely with a vendor that had experience in training government staff to defend government systems. The challenge was to develop scenarios and attack vectors that would be relevant and actionable for private-sector businesses. In this case, the defenders were from a wide variety of financial services firms, including banks, brokerages, insurance companies, payments organizations, and third-party service providers to the financial services industry. Since these cyber range exercises were held in person, we limited the number of participants but held them in numerous locations around the U.S., Europe, and Asia. Over time, we also allowed companies to participate virtually, which enabled them to have their entire cyber response teams fully engage with the exercise.

At the GRF, we have provided a number of 1-day exercises for some of the information-sharing communities we support. We have employed some major consulting firms to conduct these exercises virtually, which has been especially effective in reaching member penetration goals during the COVID pandemic. These consulting firms also have the latest technology and experienced staff to conduct relevant, interactive, and actionable exercises. As a result, we have had hundreds of participants, with most of our members participating and finding the exercises beneficial.

The ability of both the private and public sectors to share information about attacks and threats is a reason we also participate in government-sponsored exercises. This tests our playbooks for individual industry sectors to ensure that our member companies have the proven ability to respond and recover from attacks. The threat of sophisticated nation-state attacks against critical infrastructure is a primary concern.

Laurent Gerardin

After graduating from Poly'Tech engineering school (France) in 1999, Laurent Gerardin joined the BNP Paribas group in 2002, where he contributed to its cybersecurity practices, first in Paris (2 years), then in Japan (5 years), and then for the whole Asia-Pacific region from Hong Kong (5 years). In 2015, he transferred back to the head office in Paris and became the Global Head of Cyber Defense. Over 3 years, he enhanced the cyber protection of the Corporate Investment Bank. In 2018, he joined l'ANSSI (the French national cybersecurity regulator) to coordinate actions with the public and private sectors. Today, he works for APRIL, an insurance broker, as the group CISO.

These last years, with an all-time high number of *ransomware attacks*, there has been an unprecedented rise of awareness among CEOs about the utmost importance of investing in cybersecurity – but the defense has always been tougher than the attack. Ask the experts: in defending, you have to make sure that every single weakness is covered, because it will *only take one* to penetrate and destroy your business. According to the blue/red team concept, **blue teams** are the ones in charge of setting up the right (understand: *the highest cost/value*) mechanisms of protection and detection, while the red teams test them by trying hard to find the one piece of wood that will make your entire Jenga collapse.

However, if defending is such a difficult job, how to build the best blue teams? Here are a few **tips** I can provide based on my 22 years of experience in this field.

First, and this seems like an obvious piece of advice: you should hire the **right professionals**. By *right*, I do not mean *technical experts*. Technical skills can be acquired through experience and training, but *soft skills* are difficult to change or gain over time. In order to be efficient, you need people who can *collaborate*, and who will be able to face the most exhausting challenges without tearing themselves apart. In two words, you need *team spirit*. As a company leader, you can apply that rule to your entire workforce, and during times when it becomes increasingly difficult to recruit, you surely will benefit from this to attract *talent…*

Then, you have to **keep them motivated**. Money is one thing, but it is not the most relevant. You also need to keep them *trained* (cybersecurity is fast-evolving!) and give them opportunities to play with *trendy concepts and tools*. Which cybersecurity professional dreams about dealing with firewalls and antivirus, when ZTNA and EDR are the way forward? Not only will that benefit your company but it is also a way to keep your blue team *curious* and *hands-on*.

Last but not least, do not settle your blue team in an **ivory tower**, far away from operational teams. Blue teams are required to be part of the rest of IT, especially production teams. From this close relationship, they will gain *trust and respect* – hence, they will be way more efficient in the way they address problems.

And remember: cybersecurity is a *journey*, not a destination.

Peter Sheppard, BSc (Hons), MBCS, CITP, CISA

Director – Digital assurance at Business Assurance & Audit Specialists, TIAA Ltd, UK.

Peter has substantial experience in ICT auditing and cybersecurity and is the director providing leadership for TIAA's digital assurance, which includes ICT auditing, cyber assurance, digital forensics, data analytics, and information governance.

As a subject matter expert, Peter specializes in cybersecurity assurance, assisting all types of organizations in improving their preparedness for ever-evolving cyber threats. He is a member of various cybersecurity special interest groups nationally, including one for industrial control systems.

At TIAA, Peter provides strategic insight with thought leadership within digital tech and leads on the innovation of cutting-edge digital assurance services.

My perspective is slightly unusual, having provided subject matter expertise and independent assurance for the functioning of information security teams across many sectors and technologies. In my experience, I've seen some great approaches, and some perhaps *sub-optimal*. For blue teams, there are some overlooked good practices, within people, processes, and technology. Spoiler; it's not just about technology!

- Don't rely on human behavior as a defense. Mistakes happen. If you're a privileged account, it can be serious. Make sure there are layers of protection. *Defense in Depth* is not just a catchphrase.

- Staff welfare must not be overlooked during incident response. Rest is vital to ease stress and avoid exhaustion (both can lead to mistakes). Assign a *welfare champion*. Make sure staff are fed and watered!

- Training – universally wished for, but often done badly. Your attackers have tools you've never seen, skills you haven't got, and motivation. If you're not keeping your skills fresh with research, **Continuous Professional Development (CPD)**, and training, then you will fail.

- Exercise. Don't wait for a live incident to find out you're not as prepared as you should be. There are loads of great and free resources, from exercises in a box to tabletop and technical simulations. Learn from your experiences. Failure in a safe environment can be invaluable!

For challenges, my number one priority is accountability. Often, I see a single generic corporate risk – "Cybersecurity – we get hacked" – and it's given an arbitrary score. How are you supposed to treat or mitigate that risk? It cannot be owned by IT or blue team leaders. Cybersecurity needs appropriate governance and senior accountability. We have to be realistic, and, as usual, there will be financial constraints that will prevent the procurement of the latest technology solution that you *really want and need*. Therefore, if the decision to not purchase has been made outside your control, it is vital that the decision-maker now owns the cybersecurity risk, no matter how senior they are in the organization. Ciaran Martin (ex-NCSC) repeatedly stated "it's a matter of when not if" a security incident will occur. With a threat landscape changing daily, the severity of this risk must not become routine or hidden. Cybersecurity risks will always be high-likelihood and high-impact. Your skill will be treating the risk to a point it becomes tolerable, even if it only lasts a day!

Pieter Danhieux, CEO and Co-Founder, Secure Code Warrior

Pieter Danhieux is a globally recognized security expert, with over 12 years of experience as a security consultant and 8 years as a principal instructor for SANS teaching offensive techniques on how to target and assess organizations, systems, and individuals in terms of security weaknesses. In 2016, he was recognized as one of the coolest tech people in Australia (Business Insider), awarded Cyber Security Professional of the Year (the **Australian Information Security Association (AISA)***), and holds GSE, CISSP, GCIH, GCFA, GSEC, GPEN, GWAPT, and GCIA certifications.*

Beloved artist Bob Ross famously said, "*There are no mistakes, just happy accidents.*" This uplifting sentiment certainly applies to a lot of aspects of life, but if there is one place it would receive a cold reception, it's in the war room of a large corporation following a high-impact cybersecurity incident. It is a time for swift action to remediate, with little room left for understanding the deeply embedded issues that have led to compromised software. This reactive approach to security is all too common, and these days, it's a matter of *when* rather than *if* most companies will fall victim to a threat actor exploiting a mistake.

As organizations move toward more holistic security programs that encompass a hybrid approach to risk mitigation (red team attackers and blue team defenders, hopefully adopting a complementary *purple* approach of continuous improvement), it is still a fundamentally flawed strategy in the modern threat landscape. It buys into the notion that cybersecurity is little more than a tools-based process that heavily relies on automated scanning as the bastion of defense. The primary concern is that it doesn't consider the defensive power of security-skilled developers, nor include them early enough in the security strategy.

And where, exactly, does a team of secure coding superstars come from? We know their priorities are at odds with the security team, and their chief concern is building incredible features at the speed of innovation. This is why every organization must commit to nurturing their skills and changing the way we approach security from the start of software development. It takes precision, relevant training, well-fitting tools that consider a real developer's workflow, and bridging the cultural gap that exists between developers and the whole security team.

On the subject of those security-aware developers – are they red, or blue? The truth is, the best will be both. They know their code best, and awareness of both offensive and defensive security parameters is what will truly force a reduction in the code-level security issues we continue to be plagued with decades after they first reared their ugly heads. Get them hands-on with real code, understanding how attackers exploit poor coding patterns, as well as building the defensive mindset that will, over time, ensure we can banish common vulnerabilities for good.

Index

A

administrative controls 94, 106
  examples 94
adversarial threats
  likelihood of occurrence 34
adversary emulation 158
adware 62
African Vaccine Regulatory
    Forum (AVAREF) 76
After Action Report (AAR) 135
AI/ML model training 161
Amazon Web Services (AWS) 48
Application Programming
    Interfaces (APIs) 48
application security controls, considerations
  application, testing 102
  authentication 101
  authorization 101
  encryption 102
asset inventory 28-31
assurance 83-85
Australian Information Security
    Association (AISA) 174
authentication 101
authorization 101

B

baiting 67
blue team 172
  journey 164
  setting up, challenges 162, 163
  skills, requirements 10-12
blue teaming approaches
    implementation, benefits
  monitoring and surveillance 4
  recommendation to management 5
  reporting to management 5
  risk assessment 4
  security controls 4
blue team operations
  applications 44-46
  cloud 48-51
  endpoints 47, 48
  infrastructure 43, 44
  responsibilities 53, 54
  systems 46, 47
blue team, roles
  analysts 5, 6
  compliance analyst 8
  Identity and Access Management
    (IAM) administrator 8
  incident responder (IR) 6

security administrator 7
security consultants 7
threat hunter 6, 7
Bring Your Own Device (BYOD) 102
bug bounty system 9, 111
business impact analysis (BIA) 144
Business Resilience Council (BRC)
    URL 170

C

California Consumer Privacy Act (CCPA) 72
Capture-the-Flag (CTF) 12
Center for Internet Security (CIS) 19, 112
    controls 20, 21
Chief Financial Officer (CFO) 35
Chief Information Officers (CIOs) 16
Chief Information Security Officer
    (CISO) 26, 36, 41, 81
Chief Risk Officer (CRO) 36
cloud 48-51
code injection 64
command and control phase 63
Common Vulnerabilities and
    Exposures (CVEs) 5, 8
Common Weakness Enumeration (CWE) 5
compliance analyst 8
compliance requirements
    identification 79-83
compliance scanning 112
Computer Emergency Response
    Team (CERTs) 118
Computer Security Incident Response
    Team (CSIRT) 108
Configuration Management
    Database (CMDB) 4
Constant Application Vulnerability
    Scan Tools (SAST) 109

context-aware security 159
Continuous Professional
    Development (CPD) 174
corporate espionage 72
Cross-Site Scripting (XSS) 64
cryptojacking 62
cyber 55
Cyber Attack Against Payment
    Systems (CAPS) 171
cybercrimes
    impacts 72, 73
cybercriminals 71
Cyber Kill Chain 57
    actions on objective phase 64-69
    command and control phase 63
    delivery phase 62
    exploitation phase 62, 63
    installation phase 63
    reconnaissance phase 57-59
    weaponization phase 60
cyber risk management processes
    automation 22
    pitfalls, avoiding 22, 23
cybersecurity 15
    approach 73
    metricizing, considerations 15, 16
cybersecurity analyst 5
cyber security insurance 153, 155
    coverage 153
    first-party cyber coverage 154
    risk management 155
    third-party cyber insurance coverage 154
cybersecurity metrics
    selecting 19-22
cybersecurity tools 11
cyberspace 55
cyber threat actors
    corporate espionage 72
    cybercriminals 71

hacktivists 72
hobbyists 71
types 71
Cyber Threat Intelligence (CTI) 13, 117, 118
features 118
quality 118, 119
cyber threats 55- 57

D

Danhieux, Pieter 174, 175
dark web intelligence (DWI) 113
data backups 101
data encryption 101
data exfiltration techniques
baiting 67
diversionary theft 68
dumpster diving 66
impersonation 68
phishing 66
piggybacking 66
scareware 66
shoulder surfing 67
tailgating 66
watering hole 66
whaling 68
Data Layer Protection (DLP) 52
Data Loss Prevention (DLP) 108
data loss/theft attacks playbook 141
analyze 142, 143
detection 141
post-incident 144, 145
preparation 141
remediation 143, 144
triage 142
data retention 101
data security 100
data security controls 100, 101

defense
planning, against insiders 51-53
Defense Contract Management
Agency (DCMA) 26
defense-in-depth (DiD) approach 96
defense strategy 41-43
defensive AI 160
Denial-of-Service (DoS) attacks 5, 56, 64
techniques 64, 65
Department of Defense (DoD) 25
Desvernois, Anthony 169, 170
detective controls 105, 106
administrative controls 106
physical controls 106
technical/logical controls 106
types 106
detective controls, tools 113
digital forensics (DF) tools 115
SIEM tools 114
SOAR 114
threat intelligence platform (TIP) 113
Development Life Cycles (DLCs) 28
Digital Forensics and Incident
Response (DFIR) 124
digital forensics (DF) tools 115
disaster recovery
planning 149-152
Distributed Denial-of-Service
(DDoS) attacks 60, 160, 170
diversionary theft 68
dumpster diving 66
Dynamic Application Vulnerability
Scan Tools (DAST) 109

E

encryption 102
endpoint security controls 102, 103
End User Security Awareness training 4

errors and omissions (E&O) insurance 153
European Medicines Agency (EMA) 76
exploitation phase 62, 63
Extended Detection and
    Response (XDR) 160
external stakeholders 75

F

facilitator guide 135
False Positives (FPs) 11
fileless malware 62
Financial Services Information Sharing
    and Analysis Center (FS-ISAC)
  reference link 170
fire drill 152
first-party cyber coverage 154
Food and Drug Administration (FDA) 76
functional exercises 133

G

Gartner's Magic Quadrant 115
General Data Protection Law 16
General Data Protection Regulation
    (GDPR) 16, 72, 77
Gerardin, Laurent 172, 173
Global Resilience Federation (GRF)
  URL 170
governance 83-85
Governance, Risk, and Assurance
    (GRA) approach 84
Governance, Risk, and
    Compliance (GRC) 23

H

hackathons 12
hacktivists 72
hardening scans 112
hobbyist 71
HTTP flood DDoS 64
human intelligence (HUMINT) 113

I

ICMP flood 65
Identity and Access Management
    (IAM) administrator 8
impersonation attack 68
  Account Takeover (ATO) 69
  cousin domain 68
  email impersonation 68
  email spoofing 69
  MITM attacks 69
incident response (IR) 125
  life cycle 134
  narrative scenarios 135
incident response plan 131-133
  testing 133-135
incident response playbooks 136
  data loss/theft attacks playbook 141
  phishing attacks playbook 145
  ransomware attacks playbook 136
Indicator of Compromise
    (IOC) 10, 108, 120, 126
  feeds 126
indicators of threats (IOTs) 10, 113
information security championship 44
Information Security Management
    System (ISMS) 82
Information Sharing and Analysis
    Centers (ISACs) 118

Information Sharing and Analysis
    Organizations (ISAOs) 118
Information Technology (IT) asset 56
InfoSec champion 45
Infrastructure as a Service (IaaS) 48
injection attack 63
    code injection 64
    Cross-Site Scripting (XSS) 64
    LDAP injection 64
    OS command injection 64
    SQL injection 63, 64
    XML External Entities (XXE) injection 64
insiders
    defense planning against 51-53
insider threat 51, 162
    types 70
Interactive Application Vulnerability
    Scan Tools (IAST) 109
internal attacks 70, 71
internal stakeholders 75
Internet Engineering Task Force (IETF) 160
Internet Information Services (IIS) 108
Internet Relay Chat (IRC) 7
Intrusion Detection System (IDS) 30, 106
Intrusion Prevention System
    (IPS) 30, 96, 107
iterations 44

K

kanban 45
Key Performance Indicators (KPIs) 15
Key Responsibility Areas (KRAs) 4
Key Risk Indicators (KRIs) 15, 16, 77
    benefits 16
    building 77-79
    features 77, 78

Key Risk Indicators (KRIs), design phases
    baselining 17, 18
    discovery phase 17
    investigating 18
    limits 17, 18
    monitoring 18
    relevant assets, nominating 17
    reporting 18
    risk management 18
KRI collection
    automating 23, 24
KRI visualization
    automating 23, 24

L

LDAP injection 64
Lei Geral de Proteção de Dados
    Pessoais (LGPD) 16
Local Area Networks (LANs) 49
Log4j 63
logical controls 95

M

Machine Learning (ML) 160
malicious software (malware) 58
malware attacks
    adware 62
    cryptojacking 62
    fileless malware 62
    ransomware 61
    rootkits 62
    spyware 62
    trojan horses 61
    viruses 60
    worms 60

Managed Security Services Providers (MSSPs) 125

Man-in-the-Middle (MITM) attacks 59

Domain Name Server (DNS) spoofing 59

email hijacking 59

HTTPS spoofing 59

IP spoofing 59

Wi-Fi eavesdropping 59

Manufacturer Usage Description (MUD) 160

using, benefits 160

metrics 16

MITRE ATT&CK framework 127

Enterprise 127

Enterprise iteration, tactics 127, 128

ICS 127

implementing 128

Mobile 127

URL 127

MITRE ATT&CK framework implementation

cybersecurity strategy, planning 128

gaps, identifying in defenses 129

threat intelligence, integrating 129

MITRE D3FEND 129

monetization 69

multi-factor authentication (MFA) 98

N

National Health System (NHS) 61

National Institute of Standards and Technology (NIST) 25, 26, 80

implementation tiers 81

National Medical Products Administration (NMPA) 76

National Security Agency (NSA) 61

National Small Business Association (NSBA)

reference link 171

Nelson, William B 170

career 170

non-profit and volunteer work 171

network access control (NAC) 99

post-admission 100

pre-admission 100

network controls 99, 100

NIST Cyber Security Framework (CSF) 80

NIST, risk assessment methodology 27, 28

NIST, Risk Management Framework (RMF)

steps 26, 27

NoSQL attacks 64

NTP amplification 65

O

Open Source Intelligence (OSINT) 113, 118

Operating System (OS) 48, 58

Operational Resilience Framework (ORF) 170

Operational Technology Information Sharing and Analysis Center (OT-ISAC)

URL 170

operational threat intelligence 121

OS command injection 64

P

participant guide 135

penetration testing 110

perimeter security 98

personal identification numbers (PINs) 67

personally identifiable information (PII) 65, 138

phishing 58, 66

phishing attacks playbook

analyze 147

detection 146

post-incident 148, 149
preparation 145
remediation 147, 148
triage 146
phishing simulation 43
physical controls 95-99, 106
examples 95
piggybacking 66
Platforms as a service (PaaS) 48
point-of-sale (POS) 102
policy control 96, 97, 98
pretexting 58, 67
preventive control 93
administrative controls 94
benefits 93, 94
physical controls 95
technical controls 95, 96
preventive control, layers 96
application security controls 101, 102
data security controls 100, 101
endpoint security controls 102, 103
network control 99, 100
physical controls 98, 99
policy control 96, 97, 98
user security 103, 104
private cloud 49, 50
dedicated cloud 49
managed private clouds 49
public cloud 49, 50
Public Cloud Providers (PCPs) 49
purple team 9, 10

R

ransomware 61, 162
ransomware attacks playbook 136
analyze 138
detection 137
post-incident 140

preparation 136
remediation 138-140
triage 137, 138
reconnaissance phase 57-59
red team 8, 9
red team penetration testing 44
red teams 111
Remote Access Control (RAC) 48
remote access tools (RATs) 148
Remote Desktop Protocol (RDP) 162
Remote Procedure Call (RPC) 162
Return on Investment (ROI) 13
risk assessment 27
risk calculation 33-36
risk factors 32
risk management 84
responsibilities 36
**Risk Management Framework
    (RMF) 26, 164**
risk management methods
risk calculation 33-36
threat identification 31-33
risk model 32
Robotic Process Automation (RPA) 23
role-based access control (RBAC) 102
rootkits 62

S

scareware 66
script kiddie 71
scrum master 45
secret tunnel 32
secure disposal 101
security administrator 7
security awareness 33
Security Awareness Officer (SAO) 36
security consultants 7

Security Information and Event
    Management (SIEM) 23, 125
Security Operations Center (SOC) 5, 107
    benefits 108, 109
    usage 107, 108
    use cases 108
Security Operations Control
    (SOC) analysts 163
security orchestration, automation,
    and response (SOAR) tools 114
Server Message Block (SMB) 61
Server-Side Request Forgery
    (SSRF) attacks 64
Service Principle Name (SPN) 29
Sheppard, Peter 173, 174
shoulder surfing 67
single sign-on (SSO) 8, 101
SMB version 1 (SMBv1) 61
SMS phishing (smishing) 58, 69
SOC analyst 5, 6
social media intelligence (SOCINT) 113
soft controls 94
Software as a Service (SaaS) 48
software library 63
source code scan 112
spear phishing 33
sprint planning 45
spyware 62
SQL injection 63, 64
stakeholders 75, 76
    requirements 76, 77
    types 75
Standard Operating Procedure (SOP) 6
Strategic Threat Intelligence
    Feed (STIF) 119, 120
Structured Query Language (SQL) 63
subject matter experts (SMEs) 7

supply chain attacks 59, 162
SYN flood DDoS 64
systems 46

T

tabletop exercises 133
tactical threat intelligence 120, 121
tactics 127
Tactics, Techniques, and Protocols (TTP) 10
tailgating 66
talent development and retention,
    idea implementation
    Capture-the-Flag (CTF) 12
    community outreach 13
    continuous unhindered learning 13
    cyber labs 12
    hackathons 12
    mentoring 13
    research and development projects 13
technical controls 95, 96
    examples 95
tests 134
Therapeutic Goods Administration
    (TGA) 76
third-party cyber insurance coverage 154
threat 31
threat analyst 6
threat events 31
threat hunter 6, 7
threat hunting 124
    CTI, using 126
    importance 124, 125
threat identification 31-33
Threat Intelligence Analyst (TIA) 7
threat intelligence implementation 122
    analysis 123
    collection 122

dissemination 123
feedback 124
plan, developing 122
processing 123
threat intelligence platform (TIP) 113
Threat Intelligence (TI) 7, 10, 113
operational threat intelligence 121
Strategic Threat Intelligence
    Feed (STIF) 119, 120
tactical threat intelligence 120, 121
types 119
threat researcher 6
threat sources 31
types 31, 32
triaging analyst 5
trojan horses 61
Trusted Platform Module (TPM) 66

U

User Datagram Protocol (UDP)
    flood DDoS 65
user security 103, 104
US Financial Regulatory
    Framework (USFRF) 76

V

Virtual Chief Information Security
    Officer (VCISO) 159
virtual machines (VMs) 100
Virtual Private Networks (VPNs) 49
viruses 60
vishing 69
voice phishing (vishing) 58
vulnerability reward scheme 111
vulnerability testing 109-110

W

watering hole 66
weaponization phase 60
Web Application Firewall (WAF) 133
whaling 68
Wide Area Networks (WANs) 49
worms 60

Z

zero-trust 161, 162

Packt.com

Subscribe to our online digital library for full access to over 7,000 books and videos, as well as industry leading tools to help you plan your personal development and advance your career. For more information, please visit our website.

Why subscribe?

- Spend less time learning and more time coding with practical eBooks and Videos from over 4,000 industry professionals

- Improve your learning with Skill Plans built especially for you

- Get a free eBook or video every month

- Fully searchable for easy access to vital information

- Copy and paste, print, and bookmark content

Did you know that Packt offers eBook versions of every book published, with PDF and ePub files available? You can upgrade to the eBook version at packt.com and as a print book customer, you are entitled to a discount on the eBook copy. Get in touch with us at customercare@packtpub.com for more details.

At www.packt.com, you can also read a collection of free technical articles, sign up for a range of free newsletters, and receive exclusive discounts and offers on Packt books and eBooks.

Other Books You May Enjoy

If you enjoyed this book, you may be interested in these other books by Packt:

Cybersecurity Attacks – Red Team Strategies

Johann Rehberger

ISBN: 978-1-83882-886-8

- Understand the risks associated with security breaches
- Implement strategies for building an effective penetration testing team
- Map out the homefield using knowledge graphs
- Hunt credentials using indexing and other practical techniques
- Gain blue team tooling insights to enhance your red team skills
- Communicate results and influence decision makers with appropriate data

Purple Team Strategies

David Routin, Simon Thoores, Samuel Rossier

ISBN: 978-1-80107-429-2

- Learn and implement the generic purple teaming process
- Use cloud environments for assessment and automation
- Integrate cyber threat intelligence as a process
- Configure traps inside the network to detect attackers
- Improve red and blue team collaboration with existing and new tools
- Perform assessments of your existing security controls

Packt is searching for authors like you

If you're interested in becoming an author for Packt, please visit authors.packtpub.com and apply today. We have worked with thousands of developers and tech professionals, just like you, to help them share their insight with the global tech community. You can make a general application, apply for a specific hot topic that we are recruiting an author for, or submit your own idea.

Share Your Thoughts

Now you've finished *Cybersecurity Blue Team Strategies*, we'd love to hear your thoughts! Scan the QR code below to go straight to the Amazon review page for this book and share your feedback or leave a review on the site that you purchased it from.

https://packt.link/r/1-801-07247-7

Your review is important to us and the tech community and will help us make sure we're delivering excellent quality content.

Download a free PDF copy of this book

Thanks for purchasing this book!

Do you like to read on the go but are unable to carry your print books everywhere? Is your eBook purchase not compatible with the device of your choice?

Don't worry, now with every Packt book you get a DRM-free PDF version of that book at no cost.

Read anywhere, any place, on any device. Search, copy, and paste code from your favorite technical books directly into your application.

The perks don't stop there, you can get exclusive access to discounts, newsletters, and great free content in your inbox daily

Follow these simple steps to get the benefits:

1. Scan the QR code or visit the link below

https://packt.link/free-ebook/9781801072472

2. Submit your proof of purchase
3. That's it! We'll send your free PDF and other benefits to your email directly

www.ingramcontent.com/pod-product-compliance
Lightning Source LLC
Chambersburg PA
CBHW060559060326
40690CB00017B/3755